Ab

Steve Boggan has been a news, feature journalist for more than thirty years. A former chief reporter with *The Independent*, he has also written for *The Times*, *Guardian*, *Sunday Times Magazine*, *Evening Standard* and *Daily Mail*. His first book, *Follow the Money: A Month in the Life of a Ten-Dollar Bill*, was published in 2012 to widespread acclaim and was also a BBC Radio 4 Book of the Week. He lives in London.

More Praise for *Gold Fever*

'It is Boggan's wry, self-deprecating wit that makes [*Gold Fever*] so entertaining and enjoyable.'   *Daily Mail*

'A freakishly charming travel memoir... Boggan's dry, understated humour lends the just-right touch to his undertaking... An excellent writer telling a fun and interesting story.'
                                                              Library Journal

'Steve Boggan is an amazing writer who always brings to vivid life everyday people doing interesting and extraordinary things. He is somehow able to make people forget that he is a journalist and relate to him as an ordinary guy… This book is no "fool's gold", but the real thing.'
                              George Cockcroft (a.k.a. Luke Rhinehart),
                                              author of *The Dice Man*

'A well-crafted story with heady fast-forward momentum. A dogged investigator's obsessive quest for Californian gold, and the backstory of the gold fields.'
                              Iain Sinclair, author of *London Orbital*

'Steve Boggan takes the reader on an exciting adventure and proves that prospecting is as much about digging for humour and hope as it is finding gold.'

Jennifer Pharr Davis, author of
Becoming Odyssa: Adventures on the Appalachian Trail

'*Gold Fever* is a wonderful mix of history, journalism and good old-fashioned adventure. It is this last aspect that really lends the book a unique appeal; Boggan is not afraid to get his hands dirty, and he shares his experiences and findings with great depth, humour and charisma.'

Leon McCarron, author of *The Road Headed West:
A Cycling Journey Through North America*

'A beautiful book – funny, poignant, and well-written.'

Rudy Maxa, *Rudy Maxa's World*,
America's #1 travel radio show

'[Boggan] has a wonderful time, respectful of men prepared to give up everything in return for very little, but roguish enough to drop the wink on their ornery, old-fashioned optimism.'   *Saga*

'A perfect mixture of travelogue, history and down and dirty experience, leavened by rich veins of humour and pathos. A gem, or should I say a nugget.'

Mick Conefrey, author of *The Ghosts of K2* and *Everest 1953*

'Terrific. Pack your bag, grab your pick, and set out with master storyteller, Steve Boggan, for a trip in this highly original travelogue.'

Daniel Klein, author of *Every Time I Find the Meaning
of Life, They Change It* and *Travels with Epicurus*

STEVE BOGGAN

GOLD FEVER

One Man's Adventures on the
Trail of the Modern Gold Rush

ONEWORLD

A Oneworld Book

First published in North America, Great Britain and Australia by
Oneworld Publications, 2015
This paperback edition published 2016

ISBN 978-1-78074-860-3
ISBN 978-1-78074-697-5 (eBook)

Typeset by Hewer Text Ltd, Edinburgh
Printed and bound in Great Britain by Clays Ltd, St Ives plc

Oneworld Publications
10 Bloomsbury Street
London WC1B 3SR
England

For Suzanne

CONTENTS

CONTENTS

Part II

Happy Camp
300 miles

Downieville

Grass Valley

Nevada City

Colfax

Auburn

Cool

Coloma

Placerville
(Hangtown)

Sacramento

49

San Francisco

Sutter Creek

Angels Camp

Columbia

Sonora

Mariposa

49

Sierra Nevada Mountains

Lake Tahoe

N
W E
S

GOLD COUNTRY

GOLD RUSH ROUTES

Fort Hall

Fort Laramie

Fort Bridger

Sacramento City

Salt Lake City

San Francisco

Fort Kearny

KEY

—— William Swain's route
April–November 1849

---- Sarah Royce's route
1849

······ John Borthwick's sea route
1851 (inset)

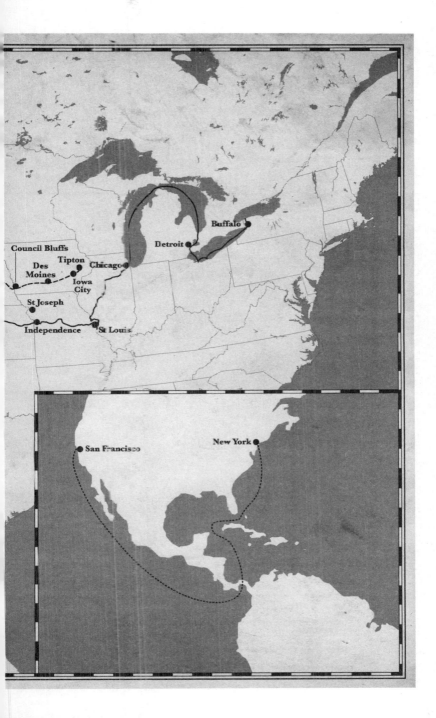

PROLOGUE

On Thursday 13 March 2008, the price of gold went above a thousand dollars an ounce for the first time in history. All over the world, at least in countries with gold-bearing soil, people with no experience of prospecting began shopping for shovels and pickaxes, gold pans, tents, generators and all manner of equipment that they had no idea how to use. And off they went mining.

Men and women whose judgement had previously gone unchallenged gave up jobs and left their homes and families in the belief that they could strike it rich. In the United States, they began to call this a new gold rush; the number of mining-claim applications rose by twenty per cent in just twelve months.

From a distance, I suspected this was nothing more than folly but I admired the people who were taking part in it and persuaded a magazine to send me to California to take a look for myself. What I found left me puzzled. The gold-seekers I met seemed broke, hungry and mostly bereft of the one thing they had come here for: gold. Yet cheerfully they trussed up all sense of reason and kept on digging.

Was this hope I was witnessing, or desperation? Because most of the stories I found were tainted with sadness, or denial, or both.

There was the couple mining for enough gold to cover the cost of their wedding; the man who had lost his job, living in a car and ignoring his wife's pleas to come home; the pensioners who had exchanged their condos in Florida for motorhomes in the hills, pretending that – heck – it wasn't really about the gold, while digging for it to the point of exhaustion; and there were gullible prospectors who were packing up and heading to a new river because someone had told them about the guy there who had hit a big pay streak.

'*The guy*' seemed everywhere and nowhere; rumours of his fantastical finds were enough to send hordes of miners rushing from one speculative camp to another. Was he a shining figment of their collective imaginations or was he real? I looked hard but I never found him.

I met a few prospectors who had unearthed fistfuls of gold but they seemed as impoverished as those who had none. After a time, they would offer to show me their hoard and all would become clear; they couldn't bring themselves to sell it. They would rather starve.

Why, I wondered, did they do this? Most admitted that they were unlikely to find the riches they had come for, but they kept on coming anyway.

I returned to London believing that there was something wrong with these people, while feeling faintly envious of them. They woke each morning to fresh mountain air, to the sound of clean, clear water rushing over rocks, in the belief that today could be their day, the day they found a rich vein or a gold nugget the size of their fist.

And the odd thing is, they were right.

I began to study the California Gold Rush of 1849 and discovered that the frantic efforts of more than three hundred thousand miners – arguably the largest single migration event the world had ever seen – had made barely a dent in the state's reserves of gold; most geologists believe that eighty per cent of it has yet to be found.

During the original Gold Rush chunks weighing as much as 160 pounds were uncovered – and remarkable pieces still turn up. As recently as July 2014 a six-pound nugget found in Butte County, California, fetched four hundred thousand dollars at auction. In an instant it had transformed the life of the man who discovered it.

Gold, I thought, was like a wish; it could change everything in the blink of an eye.

In 1849, otherwise sensible people left their families, homes and professions, risking all on fantastic reports of a new El Dorado. It sounded terribly familiar. They travelled thousands of miles over hard country, desert and mountains to get to California, to Gold Country, to the place where their dreams would come true. But many of them died – some estimates say one in five – either on their way to the goldfields or in the squalid conditions they found when they arrived there.

As I read of their adventures, I came to admire their courage and spirit. I found myself imagining where they had hunted for their gold and, increasingly, wanting to be there, wondering whether I, too, might discover some monstrous nugget, some life-changing hunk of metal.

Of course, I did nothing about this – it was a ridiculous notion, a wild fantasy.

The 2008 mini gold rush came and went before another hove into view in 2011 as gold neared two thousand dollars an ounce. It had taken six thousand years since the pharaohs first began mining gold for it to reach a thousand dollars an ounce, then just three more for that price to double. I found this utterly incredible.

And that was when my curiosity became my malady. I

promised myself that when the price of gold reached two thousand dollars an ounce, I would go and look for it. I had little practical sense of how to find it or where to look, but if I followed in the footsteps of the 1849 prospectors, the '49ers', I couldn't go far wrong.

By mid-2011, the Bank of America and Morgan Stanley were predicting that gold would crash through the two-thousand-dollar ceiling at some point in 2012. Secretly, my excitement was mounting but I was too embarrassed to tell anyone about it. They would call me a fool and how could I argue with that?

The price of gold climbed to 1,920 dollars by September 2011 and then, aside from the occasional blip, it went down and down and down. By the end of 2012, some analysts were again predicting that chaos in European economies would drive it above two thousand, but their predictions were wrong. It fell and kept on falling until by the end of June 2013 it was worth only 1,220 dollars an ounce.

The sense of loss I felt over gold's failure to reach the two-thousand-dollar mark surprised me. My excuse to go looking for it was gone and that mattered more to me than I had imagined it possibly could.

My head told me that this was a good thing; it would have been a foolish enterprise and it would have ended in tears. But my heart told me something different; that I should go. Even if I found nothing, it would be a great adventure.

You see, looking for gold is no more complicated than playing the lottery. You know the odds are stacked against you, but you buy a ticket anyway.

Before I could change my mind, I packed a bag on 1 July and boarded a flight for San Francisco, two hours west of Gold Country.

PART I

'Nature herself makes it clear that the production of gold is laborious, the guarding of it
difficult, the zest for it very great, and its use
balanced between pleasure and pain.'

Diodorus Siculus, Greek historian, first century BC

1

The Luck of James Wilson Marshall

The morning sun rose over the South Fork of the American River shortly before six, setting into relief the pointing figure of James Wilson Marshall, an ordinary man with extraordinary luck. Cast in bronze ten feet high on top of a thirty-one-foot granite plinth, Marshall looks almost embarrassed in his floppy hat and breeches, his left arm outstretched for all eternity, index finger aimed at some specific place of inscrutable interest far down below.

'Okay, okay, it was there,' he seems to be saying. 'Now can I put my arm down?'

You can't actually see the spot to which Marshall is pointing from his hilltop perch; trees and geography obscure the view. But if you've made it this far, as far as Coloma in El Dorado County, California, then you probably already know it as the exact location where, on 24 January 1848, he found gold. In a fleeting moment far less dramatic than the monument that celebrates it, Marshall simply looked down into a ditch and saw some yellow and shiny objects.

'I picked up one or two pieces and examined them attentively,' he later recalled. 'And having some general knowledge of minerals, I could not call to mind more than two which in any way resembled this: sulphuret of iron, very bright and brittle; and gold, bright yet malleable. I then tried it between two rocks and found that it could be beaten into a different shape, but not broken.'

Marshall took the nuggets to William Scott, one of his workmates, and, appropriately tongue-tied, said simply, 'I have found it.'

'What?' inquired Scott.

'Gold,' said Marshall.

'Oh, no! That can't be.'

Marshall looked into the palm of his hand. 'I know it to be nothing else,' he said.

The men were in the middle of nowhere but word seeped out and so began the most infamous gold rush in history, a frenzy of hopes born and dashed, of greed and fever, lust and death. In just eighteen months, the population of San Francisco – the nearest seaport – would grow from a few hundred to twenty-five thousand. Within two years of Marshall's discovery, California would become part of the United States, its admission a political watershed that would affect the balance of power between the pro-slavery South and the free North and eventually lead to civil war.

Over the next seven years, more than three hundred thousand chancers would pour in to slake their thirst for gold. Some of them would strike it rich; most wouldn't. And tens of thousands would die trying.

Looking up at the bronze figure of Marshall in the early morning heat – the day was already sweltering but it would reach 107 degrees by mid-morning – I began to wonder whether my timing was a bit off, perhaps by as much as two hundred years. After rushing to extract gold from the ground, I had begun to look into the practicalities of such an enterprise and learned mostly this: these days it is very hard to find gold. It is even harder to find in any significant quantity.

Before Marshall's discovery, California was a land mass almost twice the size of the UK that had been largely ignored and little-settled by the pre-eminent colonial powers of the day. It had an indigenous Native American population of around 150,000 when the Gold Rush began, a number that would be reduced by illness, starvation and murder to around thirty thousand by the time it ended.

The remaining population, numbering a few thousand, comprised Californian-born Mexicans, Americans who had survived perilous journeys from the east, Mormons escaping persecution, a sprinkling of escaped slaves and a handful of adventurous Europeans who had found favour with the Mexican authorities when the country declared independence from Spain (with California included in the territory) in 1821.

Among these Europeans was a German-Swiss adventurer named John Augustus Sutter, who had managed to persuade the Mexican governor of California to grant him almost fifty thousand acres of land in the verdant valley of the Sacramento River, a couple of hundred miles north-east of what we now call San Francisco; it had previously been called Yerba Buena. So desperate were the Mexicans to tame California that it was not unusual for them to give away large tracts of it.

Central to Sutter's empire was a settlement he had built and named after himself – Sutter's Fort – in modern-day Sacramento. It was to this fort, close to the confluence of the Sacramento and American rivers, that settlers, trappers and traders would beat a path and from which they would spread out far and wide to populate the countryside, a land he called

New Helvetia, using the Latin *Helvetia* for Switzerland, the country in which he was raised.

Sutter already employed dozens of men and owned thousands of head of livestock when, in the autumn of 1847, he commissioned James Marshall, a carpenter from New Jersey, to build him a saw mill forty miles east of the fort near the South Fork of the American River where there was particularly good lumber in the shape of enormous ponderosa pines. With this lumber would come building, expansion and power. He made Marshall a partner in the enterprise and sent him off to build the mill at Coloma in the foothills of the Sierra Nevada Mountains, with a handful of Mormons and Native Americans for labour.

So when Marshall returned to Sutter's Fort with news of the gold find a few months later, Sutter was less than pleased. He was smart enough to know that this was not good news for him; if word of the discovery leaked, his lands would be overrun by prospectors, his workers would run off to look for gold and his dreams of an agricultural and ranching empire would lie in ruins.

One can only imagine the shushing, cussing and vows of silence that were issued that cold day in January 1848 but they counted for nothing. Gold has a way of working on the tongues of men and making them slack-jawed, even when it is to their detriment. Word of the discovery passed around the small group of workers whom Marshall had employed in Coloma and so Sutter agreed to let them prospect in return for their silence. But then, as now, few men with gold in their pockets can resist boasting about it.

Within a couple of months all the settlers in the Sacramento Valley had heard of the discovery. This in itself was not enough to spark a gold rush; most were farmers who either didn't believe the reports or couldn't care less about them. Even the two weekly newspapers that existed in San Francisco, the

Californian and the *California Star*, mentioned only in passing the discovery made that winter.

It wasn't until spring that the first flames of the Gold Rush were lit, and it can be said with absolute certainty that one man and his actions – outstanding in their deviousness – fanned them.

His name was Sam Brannan, the twenty-nine-year-old proprietor of the *Star*, a Mormon who had been despatched by the movement's founders to consider California as a possible home for the Church of the Latter-Day Saints. By a remarkable stroke of good fortune, Brannan had recently established a general store at Sutter's Fort. When some of the early miners went into his shop and offered to pay for provisions in gold, he went to Coloma, acquired a jar full of flakes and nuggets from them and returned to the fort, where he stocked up with shovels, picks, canvas, pans and everything necessary for mining operations. And then he travelled with the jar of gold to San Francisco.

There he ran through the streets shouting, 'Gold! Gold! Gold from the American River!' and watched calmly as the gold he carried drove the small population into a frenzy. Having cornered the market in the region's mining equipment, he sauntered back to his store and began fleecing his gold-crazed customers. Gold pans that he had bought for twenty cents were sold for fifteen dollars, the equivalent today of 450 dollars each. Within a few short months he was the fledgling state's first millionaire.

The news brought by Brannan wreaked havoc. By mid-May it is estimated that seventy-five per cent of all the men in California had abandoned their livelihoods and homes and headed off to the goldfields of the Sierra Nevada Mountains. There were no doctors to treat the sick, no labourers to till the land and no merchants to sell food or goods. It was the same story from San Diego in the south of California to Sonoma in the north.

Before suspending publication due to the exodus of its reporters and printers, the *Californian* railed: 'The whole country from San Francisco to Los Angeles, and from the seashore to the base of the Sierra Nevada, resounds to the sordid cry of gold! GOLD! **GOLD**!!! while the field is left half planted, the house half built, and everything neglected but the manufacture of shovels and pickaxes and the means of transportation to the spot where . . . the average for all concerned is twenty dollars per diem.'

To put that into perspective, Americans 'back east' were earning about a dollar a day.

As news began to spread in newspaper reports all over the world, men – for it was almost universally men – left their ploughs, professions and families and headed to California. The first arrivals came by ship up the western edges of the continent or across the Pacific from the Orient and Australasia. Then followed miners overland from the Midwest and by sea from the east coast and Europe, braving the malarial jungles of the Isthmus of Panama or the gigantic swells of Cape Horn to reach their destination, journeys that would take between five and eight months. They came from China, Argentina, Australia, Britain, France and Ireland, from New Zealand, Russia, Mexico, Peru and Chile, from the four corners of the earth.

And when they arrived, the ships' crews left with their passengers to look for gold. By July 1849 more than five hundred abandoned vessels lay idle in the port of San Francisco. There was no one left to sail them.

Coloma

Very few people live in Coloma because it is essentially a large outdoor museum. If, as I did, you were to roll into town just after dawn, with mist rising over the grass and the American River, you would most likely see deer foraging near the road opposite reconstructions of John Sutter's saw mill and James Marshall's old shack.

If Marshall could actually see what he was pointing at from his hilltop plinth half a mile away, he would surely blush. They have given him his own park, the Marshall Gold Discovery State Historic Park, to be precise, complete with a second monument near his find, an old miner's cabin, a tea shop, some authentic jail ruins, a gun shop, a Chinese exhibit (for there were many Chinese in the 1849 Gold Rush) and a small but informative indoor museum.

I strolled along a trail on the west bank of the river to a large square stack of rocks embroidered with the words 'Sutter Mill' in yellow pebbles, close to where Marshall found his gold. It was lying in the mill race that channelled water into the wheel that powered the mill. It is an idyllic spot. The river is shallow but fast-flowing, as clear as a deer's eyes. On the far bank, perhaps a hundred yards away, the pines that attracted Marshall to Coloma still grow, rising gently to an endless blue sky. When he first arrived, the area was known as Culloma, from the Nisenan tribe's word *cullumah*, meaning 'beautiful'. It changed its named to Coloma in 1851 but it remains beautiful to this day.

I looked into the river and saw myriad shining objects that could have been gold. How wonderful, I thought, to reach into the crystalline waters and pick up a few nuggets. But I had read about fool's gold, iron pyrite that mimics the precious metal, flashing seductively from the river bed but crumbling to the touch. All that glisters, I thought.

As I walked away from the river, my concerns over timing were weighing on my mind. I had rushed to Gold Country in a chaotic and unplanned manner and I now realized that this was a mistake. The weather is almost always good in California, I had wrongly thought, and, as a person from a cold climate, too much heat is rarely a consideration. I wiped the sweat pouring down the back of my neck. It was seven o'clock on a July morning and the temperature was already in the mid-nineties.

My research had led me to a series of videos posted on the internet by a prospector named Nathaniel Burson, a young and mild-mannered individual who, unlike the prospectors of yore, did not have a wild and crazy beard, did not present himself in red long johns, unlaced boots or dog-eared hat – and who made the discovery of gold look easy. I had tracked him down to his home in Arizona and begun to seek advice by email, principally asking whether prospecting in July would kill me. He said it would not. I would be working in or near cool rivers and that would bring down the ambient temperature. But, already swooning before breakfast, I wasn't so sure.

I had a couple of major problems with which to contend: I

had no equipment and little idea of how to prospect for gold. I found myself constantly looking at my feet in the hope of stumbling across a nugget or two, lying on the ground like breadcrumbs leading the way to a huge cache. How on earth, I wondered, do you find the stuff?

Nathaniel's advice was to learn to prospect along the Klamath River in the far north of California, some three hundred miles from here, at a place called Happy Camp in the foothills of the Siskiyou Mountains. These names meant nothing to me. I called him.

'Where's the happy camp?' I asked.

'Up in the north,' he replied.

'Yes, but where's the camp?'

'Up in the north.'

'The north of what?'

'California.'

'Yes, but what town?'

'Happy Camp.'

'Yes but where's it at? In what town can I find the happy camp?'

'No, you don't understand. Happy Camp *is* a town.'

I was already learning that one of the pleasing aspects of the 1849 Gold Rush was that hitherto unnamed locations were suddenly placed on maps, no matter how rudimentary, by the first people to arrive there. Often, they would simply name a place after themselves: Angels Camp, after Henry Angel and his trading post; Downieville, after Scots gold miner William Downie; or after some devastatingly obvious feature: Dry Diggin's, Beaver Creek and so on.

Near Coloma you will find Rattlesnake Bar Road, Brush n Rocks Lane, Salmon Falls Road and Cherry Acres. Many names, such as Lou Allen Lane, suddenly become very personal and rather moving; one can imagine Mr Allen arriving, downtrodden or unlucky, perhaps, in a fresh land where anything is possible, naming this track after himself. My favourite among these was 'Myown Road' near Placerville, just ten miles or so from the statue of James Marshall.

Nevertheless, I had not expected an entire town to be called anything so ridiculous as Happy Camp. Nathaniel had told me that there were two things of interest there: good people and good gold. So in my mind's eye I saw a pine-scented encampment inhabited by cheerful, bearded old-timers who ate beans, drank whiskey and danced around crackling log fires burning a sensible distance from their dated canvas tents and the bulging sacks of gold that they housed.

Nathaniel also told me that there was nowhere to buy camping equipment in Happy Camp, a fact that left me wondering whether there wasn't a blindingly obvious opportunity to be had there. I took one last look at the American River and promised myself that I would come back and search for gold, but not as a greenhorn – as a seasoned miner, however long that might take. Then I set off to buy a tent.

I had previously camped out only twice in my life, and found it less than enjoyable each time. The first, when I was a boy, was in the back garden of my home with my best friend and was memorable for the fact that my sister brought out hot chocolate into which a large frog jumped and quickly expired. The second, many years later, was at a music festival in Wiltshire. My partner, Suzanne, and I had arrived with a new tent and began putting it up as our friends, in various states of sobriety, watched in amusement. We were delighted with the results, filled with the satisfaction that comes with the realization that no pegs, poles, pieces of vinyl or string are left over at the completion of the job. Why everyone was laughing to the point of asphyxiation we could not understand. The tent was most definitely up. It was shaped something like an igloo and was firm, safe and would surely keep us dry if the heavens opened. It took a passing stranger to solve the mystery, none of our friends having felt inclined to put us out of our misery. 'It's inside out, mate,' he said.

I went to the nearest Walmart and set about buying everything I imagined I would need. The store did not stock collapsible canoes or cleft sticks, but it seemed to have pretty much everything else. I found a tent on whose box was a picture of a happy nuclear family, gambolling with purpose around a sturdy-looking construction that appeared to be lofty enough for them to stand up in. It was forty dollars, much cheaper than the nightly cost of the motels in which I was used to staying during long trips to the United States. I was going to save some serious money here!

So I bought an inflatable mattress and a foot pump; a sleeping bag; some plastic plates and cups; stainless steel knives and forks that folded in on themselves to reveal corkscrews and gadgets that may or may not have been for removing stones from horses' hooves; a battery-powered fan; a pillow; a small lantern; a torch; a large supply of batteries; a roll of gaffer tape; some string; mosquito repellent; mosquito coils; sunscreen; antibiotic cream; sting relief cream; a first aid kit; a back support; small plastic containers for food; a large plastic container for the small plastic containers; tins of pork 'n' beans, ravioli, meatballs, corned beef, peas, carrots, sweet corn and peaches; a straw hat; a commemorative Independence Day T-shirt that read '1776 – You'll Never Keep Me Down!'; a disposable mobile phone; an international calling card; pens and notebooks; a camera (I had left mine charging at home); a memory card for the camera; four gallons of water; water-purifying tablets; a cooler chest; a vacuum coffee cup; an insulated beer-bottle holder; a collapsible garden chair; and a large knife for just three dollars, which later turned out to be every bit as useful as its price suggested.

I packed it all into the four-wheel drive vehicle I had hired, headed west towards Sacramento and on to the I-5, the road that would take me north, running almost parallel with the Sacramento River, all the way to a town called Yreka where I would join Highway 96, the route that wound and pitched westwards to the Siskiyou Mountains and my final destination, Happy Camp.

The weather was beautiful, if a little frightening in its intensity; heat that could – and, in 1849, often did – kill a person travelling unprepared and without air conditioning. Driving down from the Sierra Nevada and into the Sacramento Valley sees a brisk and remarkable change in landscape, from rugged mountains and gentle rolling hills to mile after mile of flat farmland, memorable for its lovely orchards of apples and peaches and trees heavy with nuts and olives.

By late afternoon I was more than half-way and decided to find a motel for the night. I stopped for fuel in the town of Redding and was shocked when I unlocked the car door; it felt like opening an oven to check on a joint of meat. The local radio station said it was 117 degrees and again I began to wonder how wise I had been to choose July to take on this task.

I did not stay the night in Redding, even though several circuits of its central district convinced me that it was lovely in many respects. It had a small and pretty old town that was made more appealing by the presence of the Sacramento River flowing lazily through its heart. Here, the sides of the river were steep and high, but there had been no municipal trimming and shaping, smoothing and shoring of them. Instead, fabulous houses clung to the top of the banks, which were eroding and offering casual visitors the delicious prospect – at any moment – of watching someone's home topple into the water. I spent too much time waiting for this to happen, and that might have been why all of the more attractive motels I tried to check into later were full.

So I drove on past higher, deeper and less-populated communities surrounded by white peaks that made me gasp.

Resisting the temptation to find a motel in Whiskeytown – it would have ended in tears – I stopped short at the Lakehead Lodge, appropriately located in lovely isolation at the head of Shasta Lake, the green and serene run-off from the magnificent Mount Shasta, which stands almost exactly half as high as Mount Everest, and took a room for the night.

The next morning was crisp, blue-skied and full of promise. I was so close to gold I felt I could touch it. From the outset, I had decided I would assess myself daily for signs of gold fever, in much the same way as a doctor might check the temperature of someone teetering on the brink of a serious illness. Aside from leaving behind my partner, abandoning my regular work and travelling half-way around the world, I had done nothing yet to suggest I was becoming obsessed with the precious metal. The *Chambers Dictionary* definition of gold fever is 'A mania for seeking gold', and I most certainly did not have that.

In the weeks leading up to my decision to come to California, I had immersed myself in the study of gold, yet felt singularly unqualified to find it. I discovered that the element's chemical symbol was Au, derived from *aurum*, its Latin name, which means 'shining dawn' (Aurora was the Roman goddess of dawn). I learned that gold was unfeasibly dense, a cubic foot of it weighing half a ton. And that it was almost as soft as putty, an ounce of it so ductile it could be teased into a thread fifty miles long. If you had a hammer and an inordinate amount of time in which to put it to a seemingly pointless task, you could beat an ounce of gold into a gossamer-like sheet that would cover an area of one hundred square feet at a thickness of 280-thousandths of an inch.

Gold is incredibly rare – all of it found so far, about 190,000 tons, could be carried in a medium-sized oil tanker. Think of it: all the gold rings and chains, coins and bars, goblets and plates, thrones and crowns, swords and daggers, statuettes and ornaments, oddities, trinkets, bespoke adornments and teeth by the hundred million, jealously guarded by the entire human race, could fit into the hold of one ship as a cube measuring sixty-seven feet by sixty-seven feet.

Gold is inert and timeless, one of the reasons that it is associated with immortality; it never corrodes or ages. You can find gold jewellery, coins and exhibits in museums all over the world that might have been mined five thousand years ago but that look – and will always look – as if they were crafted yesterday.

Wars have been waged, lands lost, power forged, hearts broken, dreams shattered, peoples enslaved, fortunes made and lost, morals eroded with alarming alacrity, and mythology given wings by gold and our obsession with it. Gold's legitimacy and importance among Judeo-Christians comes from the Bible, in which there are more than four hundred references to it, not least in Exodus, in which God supposedly instructed Moses to construct a sanctuary in which the Jews were to worship him, together with a tabernacle to be housed inside it. God apparently said: 'Thou shalt overlay [the tabernacle] with pure gold, within and without shalt thou overlay it, and shalt make upon it a crown of gold round about.'

Wasn't this the same God whose son said it was easier for a camel to pass through the eye of a needle than for a rich man to enter the Kingdom of Heaven?

The financial historian Peter L. Bernstein, in his seminal book *The Power of Gold: The History of an Obsession*, describes John Ruskin, the art critic and social thinker, musing on the subject more than a hundred years ago. Ruskin told the story of a man travelling on board a ship with all his wealth in the form of gold coins. After a few days, a violent storm rose up and passengers were ordered to abandon ship. The man strapped his bag of coins to his torso, jumped overboard and sank into the depths. 'Now,' asked Ruskin. 'As he was sinking, had he the gold? Or had the gold him?'

I had no gold, but already I was wondering whether it had me.

My studies had led to my being haunted by figures from the past, figures in whose footsteps I was about to tread. Those who braved the California Gold Rush called themselves Argonauts after the mythical crew of the Argo in which Jason went in search of the Golden Fleece, and many of them were prolific diarists. As I read of their adventures, three of them touched me most profoundly; they were John 'J.D.' Borthwick, Sarah Royce and William Swain. Of the hundreds of accounts that have survived, I found theirs most poignant and powerful. I began to think of them as companions.

Borthwick, a Scots artist and son of an Edinburgh physician, had been born in 1824 with perpetually itchy feet. He caught the gold bug in 1851 following periods of travel through Canada and as far south as New Orleans in the United States. In a subsequent memoir, *Gold Rush: Three Years in California,* Borthwick described his adventures; he seemed one minute cavalier and brave, the next compassionate and funny. I was to tailor my prospecting just to see where Borthwick went.

My second companion, Sarah Royce, was English, having been born in Stratford-upon-Avon. She was carried to Rochester, New York, by her parents when she was just six weeks old. When her husband, Josiah, also English-born, read the first vague accounts of a gold rush far away, he announced that he, Sarah and their two-year-old daughter, Mary, were to leave their home in Iowa on the last day of April 1849 and head to California.

Very few men took their families along and this meant that Sarah, then in her early twenties, was to become one of the first women in the Californian goldfields. Her courage and resilience were as frighteningly impressive as her determination in the face of death. She instilled in me a sense of the horrific privations endured by the Argonauts during the six-month overland journey to California and of the hardships that they endured when they finally arrived there.

My third friend, William Swain, was a twenty-seven-year-old farmer from Youngstown in the northern part of New York State. An upright and honest man, he was decency personified. Recently married and with a new-born baby girl, William nevertheless had a thirst for adventure and, in common with many of those who rushed to California, he believed a year's hard work there might provide enough riches to allow his family to live out the rest of their days in comfort. Luckily, William was the keenest of diarists, an intelligent and resourceful man whose account of the journey west is regarded as the most complete of any of the Argonauts.

Once they had reached out to me from the past, I was to feel their presence in every step I took, in all the ditches I dug, in the many riverbeds I plundered.

3

Yreka, the Dredging Thing and Happy Camp

North of Mount Shasta the landscape changed dramatically into flat, yellow-grassed prairies, morphing from rugged peaks into the Shasta Valley 2,500 feet above sea level, a broad and fertile plain that must have seemed like paradise to early settlers braving the trail west and then south from Oregon. Here, in March 1851, a mule-train packer named Abraham Thompson found gold near a spot already bearing the typically descriptive name of Rocky Gulch. Within a month, two thousand miners had flooded south from Oregon and north from diggings that had already played out in the southerly Californian goldfields.

I was on my way to meet a miner named Rich Krimm, one of the 'good people' from Happy Camp who Nathaniel suggested might help me. After just two phone calls, Rich had offered to take me under his wing, but he warned that my lessons in finding gold would not begin anywhere so prosaic as a river. Instead, he had asked me to meet him in

Yreka, some eighty miles short of my final destination, at the Siskiyou County Superior Courthouse. I didn't ask why; I naturally assumed it would be something to do with gold, whiskey and guns.

I had first imagined that Yreka was an adulteration of the word *eureka*, the Greek for 'I have found it' (and the motto of the State of California), but I was wrong. Although its exact origins remain unclear, the most likely derivation can be found in the Shasta Native American word *wáik'a*, meaning 'white mountain'. However, in his autobiography, Mark Twain, who traveled extensively through gold era California and wrote entertainingly about it, took a different view. In common with all Gold Rush towns, Yreka began as little more than a tented city struggling to cope as humanity poured in. Twain wrote that a baker had arrived in the settlement and stretched out to dry a painted canvas sign bearing the word BAKERY, which had twisted in the wind to hide the letter 'B'.

A stranger came along, saw the sign from behind and thought he had arrived in YREKA. And, according to Twain, the name stuck. Today, the city has a population of about eight thousand people, wrapped around an old town that looks like a western movie set; much of the original Gold Rush architecture survives.

The Superior Courthouse is a beautiful art deco edifice strangely at odds with the surroundings in which it stands. It has a swagger that was only slightly diminished in my eyes by the fact that it had once proudly housed the town's collection of gold, a 1.3 million dollar cache of nuggets and jewellery, until it was stolen from its display case in January 2012 when someone left a window open. The fact that the biggest crime in Yreka's history had taken place in the courthouse was something I decided not to dwell upon when I met its townsfolk.

On the first floor was a motley group of individuals who

looked as if they didn't belong. Some of them looked as if they didn't belong anywhere. There was much grumbling and huffing, and one or two were moving their shoulders in a circular motion as if trying to remove the darned shirt that they had felt obliged to wear that day. They were gold miners and from their midst came a man I first took to be their lawyer; he was tall and smart, handsome, greying and incongruously well-groomed behind a magnificent moustache.

This was Rich Krimm. Rich was aged sixty-five but was in much better shape than me or anyone else in the courthouse. He approached me and crushed my hand while I smiled back at him. He was not a lawyer, but the Internal Affairs director of the New 49er Prospecting Association, a private company that controlled more than a hundred miles of mining claims along rivers in the region.

That's right: claims. The gunfights you saw on television as a kid because Red Buttons had jumped Audie Murphy's claim were over exactly the same kind that still exist in law today in nineteen of America's fifty states. If you are American and can find a piece of public land that has no claim on it, and you uncover some gold there, you can mosey on down to your nearest Bureau of Land Management (BLM) office and, under the General Mining Act of 1872, register a claim. The BLM administers almost one-eighth of the land mass of the USA.

If you're going to dig an underground mine to look for gold – what's called a 'lode' mine (a lode is a subterranean vein of metal-bearing ore) – your claim could cover an area up to fifteen hundred feet in length and six hundred in width. If you are looking for gold near the surface, usually in or near a river – what's called 'placer' mining (pronounced 'plasser') – then your claim would cover up to twenty acres – almost nine hundred thousand square feet. And each of up to eight partners could add a similar amount of land to your collective claim.

That's an awful lot of land and suddenly all the metals and

minerals on or in it belong to you. You hop (possibly skip and jump) into the BLM with a map of your claim, a 34-dollar 'location fee', a 194-dollar 'filing fee', a 140-dollar 'maintenance fee' and the claim is yours to work. You must also register it at the County Records Office, and once you've done that you're a potential millionaire or a fool with an awful lot of worthless land to dig up.

You'll notice there are quite a lot of fiddly details involved in all this, including fees to be paid, forms to be filled out and maps to be drawn. You also have to build a three-foot stack of stones or stick a pole in the middle of your claim – I am not making this up. You must also renew your claim each year and, if you forget to do this before 1 September, it will expire, which is exactly what your sneaky and avaricious neighbour is hoping for.

Since the 1980s, Rich's boss, a legendary prospector named Dave McCracken, known to miners as Dave Mack (not Dave or Mack, but always Dave Mack), had been assiduously gathering claims one by one, mostly along the Klamath River, and now he was operating a considerable business charging prospectors to look for gold on those claims. It was the job of Rich, a former police officer, to keep claim jumpers and non-payers off the land and to make sure that miners hunted for their gold legally.

Eventually, curiosity got the better of me. 'Why are we here?' I asked.

'Because of the dredging thing,' replied Rich.

'The dredging thing?'

Rich explained that one of the best ways to find gold was to climb into a river with what can only be described as a huge vacuum cleaner mounted on a floating platform and to suck up the sediment as close as you could get to bedrock. This is called dredging. Because gold is nineteen times heavier than water and around seven times heavier than anything else likely to be found in it, that is where you will uncover gold: at the bottom. It's gravity, pure and simple, and with a suction dredge you can process large amounts of sediment, filter out the stuff you don't want and be left with 'pay dirt', the mud and sand in which you are most likely to find flakes and nuggets.

In the distant past, when prospecting was carried out without regard for the environment, vast dredges as big as ships would tear up river beds, spewing out unwanted mud as they went along, polluting waterways and destroying the habitats of freshwater species. This kind of dredging had been banned long ago; today's dredges are little more than rafts carrying pumps that suck up sediment and provide air to miners working on the riverbeds below.

In recent years, environmental groups had been arguing that suction dredging was bad for rivers, in particular for the spawning grounds of salmon. As a result, the State of California had banned it. The miners argued that this was ridiculous; they weren't allowed to dredge during the spawning season and, anyway, the disruption that their vacuum hoses created was nothing when compared with the raging floods caused by springtime thaws and heavy rains each year.

The ban had crippled small mining towns that relied on the money brought in by prospectors. Each side, entrenched in the veracity of its argument, had expert opinion and environmental reports to support it.

The New 49ers and the local businesses in Siskiyou County had formed an alliance to contest the ban, based on a rather flimsy technicality. The state had defined a suction dredge as a machine with a pump attached to a sluice box, a metal tray that filtered away unwanted surface sediment. To get around this definition, the 49ers had simply removed their sluice boxes and renamed their remodelled dredges as 'gravel transfer systems'. The environmentalists weren't buying it.

I had stumbled into a pivotal moment in the argument. A

judge was going to rule whether the miners could continue with their so-called 'gravel transfer systems' in Siskiyou County.

We all settled down while the New 49ers' lawyer outlined their argument. Then, by phone and played over loudspeakers in the courtroom, a high-powered attorney for the environmentalists responded from his office in San Bernardino, more than six hundred miles to the south. I might have been wrong, but I felt he could have been filing his nails and ordering lunch while he presented his case to the hick judge from up north. The judge, however, was rather magisterial and in complete control of her courtroom. She granted the miners a thirty-day order allowing them to use their customized mining platforms in the county until another hearing, in a court in San Bernardino, would settle the matter once and for all (though at the time of writing it still hasn't been).

The highlight of the afternoon came when the judge read out the wording of her order. Over the loudspeakers, the high-powered lawyer challenged her on the placing of two commas. Everyone held their breath and looked at the judge. She didn't flinch.

'Now listen here, sonny,' she said. 'My husband is an English language teacher and I don't even let him correct my grammar!' And she didn't change a thing. The gold miners exploded into wild applause.

Okay, she didn't call him 'sonny'. But she should have.

After the hearing I followed Rich west on Highway 96 towards Happy Camp and began to feel we were venturing truly into the wild. The road wound and crawled, twisted and coiled through towering mountains, slavishly hugging the deep and slow, shallow and white waters of the Klamath.

Rich had never considered looking for gold until he had retired from his job as a police officer in 2003. He was introduced to prospecting by his brother-in-law and was hooked the first time he saw gold in a pan. He met Dave Mack at a mining fair in Las Vegas a short time later, bought a New 49ers membership and was now one of Dave Mack's dredging partners, taking underwater mining to new levels, diving into ridiculously fast-flowing water and finding ounces of gold at a time. They called it 'extreme prospecting' and it was incredibly dangerous. He would tell me all about it, he said, but first I would have to find somewhere to pitch my tent in Happy Camp.

We reached our destination shortly after six in the evening. I climbed out of the car and found myself spinning around, awed by distant beauty in every direction. The town was surrounded by the Marble Mountains, Desolation Peak, Rattlesnake Peak and Preston Peak, at over 7,300 feet the highest in the Siskiyou Wilderness. Later, I was to buy a detailed map of the region and whole swaths were labelled just that, 'Wilderness', a legal designation reserved only for America's most beautiful locations, countryside that would come under federal protection forever. I could not have been more delighted had there been descriptions that read 'Here Be Dragons'.

At the foot of the mountains, nestled among forests of ponderosa and sugar pine, madrone, spruce and Douglas fir, was Happy Camp itself, home to a population of just twelve hundred. There was a broad open lot at the town's centre with a grocery store called Parry's Market, an auto-repair shop, a tourism office, a gift shop, one motel and a camper van park. There was one restaurant, a cafe, a saloon, a variety of Karuk Native American tribal offices, a forest service centre, a self-service gas station, the Double J liquor store, a branch of the Scott

Valley Bank, a clinic and the headquarters of the New 49ers –
a brick building whose windows (tellingly) were obscured by
security grilles.

Set back from Highway 96 were pretty streets laid out on a
grid, while on the edges of town they hugged the creeks and
streams that fell from the mountains. Most homes were
wooden, well-tended and neat, though a handful appeared
rather frightening, with old heaps of twisted metal and canni-
balized cars in the yard, occasionally with advertising hoardings
outside that read 'For Rent: Movie Set – Deliverance/ Texas
Chainsaw Massacre/ Psycho – All Catered For. Children
Welcome.'

Happy Camp's most dominant feature was a twenty-foot-
high, raggedy and quite beautiful metal sculpture of Big Foot
by the artists Cheryl Wainwright and Ralph Starritt. Big Foot
was said to frequent the area. People claimed they had seen
him ambling across the road in the middle of the night, wide
old eyes shining red and a look of terrible boredom on his face.

Rich disappeared for a few minutes while I took in the view.
When he came back, he led me into the Big Foot RV (recre-
ational vehicle) Park, where he introduced me to a shirtless,
pot-bellied man who bore a striking resemblance to Edward G.
Robinson. His name was Terry McClure and he did not seem
particularly pleased to see me. He was living there with his
wife, JoAnne, in a large camper van and he had been dra-
gooned into looking after me.

'This is Steve,' Rich told Terry. 'He's come looking for
gold.'

Terry rolled his eyes.

'You can put your tent over there,' Terry said, pointing to a
spot of grass next to a dilapidated camper van. 'Shouldn't get
into too much trouble.'

Rich was in a hurry to get away but said he would come and
find me the next day. He planned to take me twenty miles

downriver to a place named Independence Camp, where there was a group of miners trying to eke out a living. 'They're a little rough and they like to drink a lot at night, but, boy, do they work hard,' he said. 'They're up at five every morning and they get straight down to it, working through the worst of the heat. I think you'll like them. They'll find you . . . interesting.'

When he left, an amused smile on his face, there was an uncomfortable silence. 'I'm a complete greenhorn,' I stuttered, filling it inappropriately. Terry frowned. He was sixty-nine years old and a seasoned prospector. 'Really, when it comes to looking for gold, I haven't a clue. . .'

My attempts at small talk tailed off. Before my arrival Terry had been enjoying a pleasant evening outside the big white caravan he called home and the large truck that pulled it whenever he and JoAnne decided to move on. They had a small garden, a shed, an awning to shelter them from the blazing sun, a barbecue and a pen from which a handful of small dogs yapped at me.

The evening was hot and sticky and I was covered in sweat before I had removed my tent from its box. I promised Terry that I wouldn't bother him and I kept my word for a whole minute before asking if he had a mallet I could borrow to hammer in my tent pegs. He fetched one from a big tool box, poured himself a cup of coffee and leaned on his truck while I got to work. It was some time before I realized that he wasn't there to enjoy the difficulties that I had anticipated, but to look out for me should they occur.

After a few minutes a greying man emerged from the battered camper van next to the spot where my tent was taking shape. 'You need some help?' he asked, holding out his hand. His name was Tom Henderson, a Scot who lived in Australia but who had come to California to strike it rich. I was later to find out two things about Tom: he was the nicest person in Happy Camp and probably the unluckiest.

Disturbed by the growing chatter, a diminutive, white-haired woman arrived holding a chihuahua and asked what all the fuss was about. Her name was Marcy Darby, she was seventy-four years old and she lived in a flower-covered RV next door to Tom.

'Are you from Australia too?' she asked. 'Tom's from Australia.'

'I keep telling you, I'm from Scotland,' said Tom with mock exasperation.

'And I'm from England,' I said.

'Is Scotland part of England?' asked Marcy.

'No!' we shouted in unison.

We chuckled while Marcy wandered off, shaking her head. Tom and I were to become great friends, spending long and happy evenings together outside my tent, planning ways to find gold. Not only did Tom suffer from bad luck, but also he endured poor health; he had had to cut short a diving career due to late-onset asthma and he was clinically addicted to oxycodone, a synthetic form of morphine, prescribed for a serious back injury.

I didn't have a table for us to sit around but Tom soon rustled up an upturned float from his dredge and before long we were joined there by Craig Harger, a thin, muscular drifter from Sacramento who looked a full fifteen years younger than his forty-eight. He came bearing cold beer. Tom did not drink or stay up late because of his medication, so it was Craig with whom I would sit into the small hours every night discussing his crazy and brilliant ideas for inventions that would never come to fruition. He had a fierce intellect but a complete lack of control over it. As a child, he had been labelled 'troubled . . . bipolar . . . mildly psychotic' by adults who didn't know how to harness his restlessness; by the age of seven he had been arrested for assault and battery.

Craig was once on the run for five years in the misplaced

belief that he was wanted by the police. That had taken him from age nineteen to twenty-four; a motor bike accident and a broken back had occupied him for the past two years; marriage, bereavement and restlessness had consumed the rest. Every day I looked forward to my late-night sittings with Craig, wondering where our conversations might take us, even if I did feel he was struggling to harness that huge brain of his.

It was still warm when everyone said goodnight and went to bed. My tent seemed to be the only one on the site and I began to wonder whether I might not be run over by some monstrous RV at dawn, but I was too tired to care.

The isolation of Happy Camp meant that its sky was clear and unsullied by light from other towns or cities. I saw a shooting star before turning in and made a wish so obvious as to be embarrassing.

4

The Island of
California

If you have ever wondered why Americans regard themselves
as lucky, consider this: California, gold and all, was ceded to
the United States by Mexico on 2 February 1848. Americans
and Mexicans had been at war for two years following the US
annexation of Texas, the Americans emerging as victors. The
Treaty of Guadalupe Hidalgo, which ended hostilities, granted
the US everything west of Texas and north to Oregon in
exchange for fifteen million dollars. It also advanced what
Americans had for some time been calling their Manifest
Destiny – the God-given right to expand, to settle and to introduce democratic values from coast to coast.

While completing the unification of a vast land mass with
beaches pounded by waves from the Atlantic and Pacific oceans
for just fifteen million dollars might be regarded as 'lucky',
what I am actually trying to draw your attention to here is the
date: 2 February 1848. That is just nine days after James
Marshall found his gold, and a couple of months before anyone

outside Coloma and the Sacramento Valley – including the Mexicans – knew anything about it.

What is strangely odd about all this – creating an eerie impression that destiny did, indeed, lend a hand – is that the existence of a place called California, a place awash with gold, had been foretold as long ago as 1510 in a Spanish novel called *Las Sergaz de Esplandián* (*The Adventures of Esplandián*) by Garci Rodriguez de Montalvo.

De Montalvo's mythical country is an island ruled by a queen called Calafia, where Amazonian goddesses fight with golden weapons and which 'abounds with gold and precious stones and upon [which] no other metal is found.' California, of course, is not an island, but for sixteenth-century Spanish explorers coming from the east, it might as well have been. Reaching it would have involved crossing the Atlantic Ocean to the east coast and then traversing the vast expanse of what is today the contiguous USA, including the Great Basin Desert and the Sierra Nevada mountain range to the east of California. Or negotiating the Mojave and Sonoran deserts to the south. Or heading north through the jungles and mountains of South America (once, as the conquistador Hernán Cortés did between 1519 and 1521, you had conquered the Aztecs). Or sailing around Cape Horn, which, in your tiny wooden vessel, and without accurate charts or maps, would have been almost suicidal.

After Cortés had defeated the Aztecs he wrote of an island to the north-west of Mexico that was probably the Baja California peninsula, the long appendage drooping parallel with northern Mexico from American California, or Alta California as the Spanish called it. Nevertheless, a land called California had become a reality. However, early explorers soon came to the conclusion that it was of little interest. A Portuguese expedition led by Juan Rodriguez Cabrillo in 1542 surveyed approximately four hundred miles of coastline, from what we now call San Diego to Point Reyes, about thirty miles north of San Francisco, and found neither the treasure nor the civilizations he had hoped for.

No one bothered much with California after that. Sir Francis Drake stopped by in 1579, claiming it for England and naming it Nova Albion. On finding only small tribes of Native Americans who had never heard of gold, he climbed back aboard the *Golden Hind* and continued westward on his circumnavigation of the globe.

In 1602 the explorer Sebastián Vizcaíno mapped large sections of coastline, naming features such as San Diego Bay, Santa Catalina Island and Monterey Bay. But his maps were not used for another 167 years, until the Spanish viceroy in Mexico City sent out an expedition of sailors, soldiers and a handful of priests to colonize the region. This involved mostly settling missions tasked with introducing the Native Americans to Christianity. Few other people ventured there from Mexico or Europe because, in reality, California was not seen as a desirable place to live. So resistant were people to going there that the Mexican government shipped convicts and outlaws to settle it.

After Mexico had declared independence from Spain in 1821, it began a policy of reducing the influence of the missions and redistributing their lands, giving away huge tracts in the hope of encouraging real colonists to develop them.

John Sutter, the man who commissioned James Marshall to build the saw mill in Coloma, was one of these beneficiaries. He had fled from debtors in Switzerland, leaving behind a wife and family, and arrived in California in 1839 via New York, Missouri, New Mexico, Oregon, Hawaii and Alaska. Once there, he ingratiated himself with the Spanish governor of California, lying about having been a captain in the Swiss Army (he had been an under-lieutenant in the Swiss reserves), and was appointed a captain of the militia, a mob that later proved itself worthless when confronted by real American soldiers. Identifying him as exactly the sort of man needed to help grow California into something useful, the Mexicans gave Sutter almost fifty thousand acres of land and appointed him an *alcalde*, a sort of magistrate with mayoral powers. In return, he was to protect the land and the Spanish-speaking population – the Californios – who lived on it.

The future for Sutter and Mexican California, then, was looking up. But not for long.

To say their fortunes were spoilt by the war with the United States and the discovery of gold would be an understatement. Imagine fleeing from debts in one country and securing vast amounts of land and potential wealth in another, only to find yourself in the middle of a gold rush that would see it all overrun. And imagine surrendering a piece of territory for a relative pittance only to find out that unimaginable wealth had been found there just nine days earlier and you didn't know because telephones hadn't been invented.

The total amount of gold estimated to have been mined in California to date is about 120 million ounces. At the time of writing, the price of gold is around thirteen hundred dollars an ounce. That represents a loss to Mexico of about 156 billion dollars at today's value, taking into account the fact that the fifteen million it received for California in 1848 would be worth about 440 million today. (I readily admit that this is a highly simplistic view of the sums involved; for many years, as we shall see, the US government pegged the price of gold at twenty dollars sixty-seven cents an ounce and, later, at thirty-five dollars. But you get the picture.)

It seems incredible that for more than three hundred years no one had thought to properly explore and survey the foothills of the Sierra Nevada, but they simply hadn't. There had been a small find in 1842 in the Santa Clarita Valley in modern-day Los Angeles County, but not enough to generate a full-blown gold rush.

Of course, you might argue that the Mexicans had lost the war and couldn't have kept California anyway. Perhaps so. But would they have fought harder, fielded more troops, bought more weapons, called for assistance from other countries (Perfidious Albion, anyone?), if they had known how much gold was lying around and about the place?

They had already seen their indigenous forbears, the Aztecs, and their Peruvian neighbours, the Incas, lose fortunes in gold and silver to an occupying power, the Spanish (though the Spanish had seen this too – the Romans plundered much of their gold in the first century BC). Cortés in Mexico, and later Francisco Pizarro in Peru, destroyed their cultures and enslaved their peoples, emptying the hillsides of precious metals. In one month alone in 1532 – after Pizarro had taken hostage the Inca ruler Atahualpa – the Spaniards extracted seven tons of gold as ransom. The conquistadors killed Atahualpa anyway.

During the next hundred years, the amount of gold circulating in Europe increased fivefold. Spanish maritime records suggest that 160 tons of New World gold eventually found its way into the country's vaults (and many times more silver). However, gold can be as much a curse as a blessing. The arrival of so much of it in Spain had disastrous consequences, causing rampant inflation and playing havoc with the country's economy. Instead of investing the wealth in domestic enterprises, Spain spent it in other European countries, making them stronger while Spain became ever more reliant upon their exports.

At the end of the sixteenth century, Cortés wrote: 'Although our kingdom could be the richest in the world for the abundance of gold and silver that have come into it and continue to come in from the Indies [Central and South America], it ends up as the poorest because it serves as a bridge across which gold and silver pass to other kingdoms that are our enemies.'

Should Mexico take some comfort from this? Would Mexico

have made better use of Californian gold than Spain made of its gold from the New World? We will never know.

I woke from a troubled, damp sleep on my first morning in Happy Camp. During the hours of darkness, the temperature had fallen to just fifty degrees and I had spent the night shivering in my tent. The sun and temperature were rising again, but my joints were creaking and I felt very old. I wanted to buy some equipment and begin prospecting straight away; all I needed to know was how.

Terry emerged from his RV and saw me sitting disconsolately in the crumpled clothes that I had put on in the middle of the night.

'You want some breakfast?' he asked.

Terry was a retired pipe fitter from Seattle, Washington; JoAnne a 'career mom' from Winnipeg in Canada. She was sparky and quick and squinted beneath her spectacles, more because of her generous smile than from the sun.

'We've been married before,' she said. 'He's my second; I'm his last.' She was sixty-eight and almost permanently cheerful. Between them Terry and JoAnne had seven grown-up children, thirteen grandchildren and three great-grandchildren who all thought the couple were insane when, fifteen years earlier, they had decided to sell their property, buy a mobile home and spend their time following sunshine and looking for gold and gems.

'I'd always wanted to go prospecting for gold but I didn't get the chance until I met Terry,' said JoAnne. I was nodding politely between mouthfuls of bacon and egg. 'Some people we knew took us out prospecting in the Arizona desert twelve, thirteen years ago and that was it. We were hooked.'

So hooked were they that for three years Terry took a job at a mining store in Quartzsite, La Paz County, Arizona, a tiny but strangely magnetic desert town that attracts huge numbers of people interested in looking for gold and precious stones. Every January and February, prospectors and amateur mineralogists such as Terry and JoAnne swell the regular population of 3,500 to more than a million. Here they trade and talk, drink and sing in the winter warmth. Around ten years ago, the couple discovered Happy Camp. Terry walked into the New 49ers store and ended up working there. He was now retired.

'We usually spend the winters in Quartzsite and the summers here looking for gold,' he said. 'It's a wonderful way of life. Our kids were worried about us at first; they thought we were crazy to just sell up and move out like that.'

You might understand their children's consternation when I tell you that the trailer in which Terry and JoAnne lived cost 110,000 dollars and the truck that pulled it a further fifty-five thousand. But it was bright and spacious inside, with all the comforts of home. The couple spent their days prospecting and their evenings walking, taking meals outside or just cosying up in front of the TV. Terry enjoyed reading; JoAnne made jewellery from the gems she found, bought or swapped in Quartzsite. It seemed to me that they spent much of their time just being happy.

'How do you find gold?' I asked.

I had been wondering for some time how to broach the subject without sounding completely naive or stupid, but I came to the conclusion that this wasn't possible. I had two options: I could head off on my own to God-knows-where to look for signs of God-knows-what without suffering the indignity of asking seasoned miners how to find gold, or I could hold up my hands and admit to being the innocent I was. The fact that the people I had met so far were nowhere near as abrasive or secretive as I had expected helped me to arrive at my decision.

Terry stood up and went into the trailer, emerging with an aluminium case. He squatted next to me and opened it to reveal some of his gold. There were nuggets and vials of flakes, twisted pieces of shining permanence that looked as if they had been frozen mid-flow on the downside of a prehistoric eruption. Some were tear-shaped, others like flattened brass. Some had the appearance of liquid, suspended. I felt a flutter in my stomach.

'I like to keep some close by but the rest is in a vault,' he said. 'Safer than the dollar. Gold will never be worthless so it's comforting to know it's there. We don't need to spend what we find right now, but who knows what the future will bring?'

'But how do you find it?'

'Well, there's lots of ways, but they all have one thing in common: hard work. You gotta move earth or sand, gravel or rocks to get to gold, because it's always on the bottom. You gotta do it in cold and heat and you gotta put your back into it, either on land or under the water. For the small miner there aren't too many shortcuts.

'There's lots of people who think they can just turn up and the gold will find them. There was this one guy we used to call "Ounce-a-Day". He was a truck driver who quit his job so he could go gold prospecting but he was the laziest sonofabitch I ever met in my life. He turned up with a one-week plan to learn how to prospect and we did our best to teach him. But we were working on the other side of the river and every day we had to haul all this equipment to our spot and ferry it over – sluice boxes, motors, pumps, weights for diving, hoses – and we'd turn round and he would be carrying a trowel or an empty bucket. That's why we called him "Ounce a Day". Not for what he found, but for what he carried. He went home after a while and didn't come back.

'There was this other guy came into the 49ers store one day. He had a big RV, a four-wheeler [quad bike], and was bragging about all the best brand new equipment he had, and he walks up to me and says, "Where's the gold?" And I says to him, "There's seventy thousand acres out there. Go find it yourself. If I knew where it was, you think I'd be working in this store?"'

'I'm not afraid of hard work,' I interrupted.

Terry and JoAnne looked me up and down.

'You got any equipment?' asked Terry.

'I've got something that might take stones out of horses' hooves,' I said, staring at my feet.

The couple looked at one another and it occurred to me that they had probably seen every type of prospector come and go over the years: the dreamers, the fevered, the braggarts, wastrels and drifters. They had learned not to waste their time on lost causes but, it seemed, they were prepared to waste it on me. They smiled.

'Come on,' said Terry. 'Let's get you some.'

He took me to the store at the New 49ers HQ and had to forcibly pull me away from a display cabinet inside the doorway in which was a bowl containing a small pile of gold nuggets, flakes and pickers (pieces that could literally be picked up between fingertips). I couldn't take my eyes off them.

'That's Dave Mack's gold so far this year,' said a woman from behind the counter. Her name was Montine and, in conjunction with her colleague, Myrna, she dealt with all the miners as they came into the store (most prospectors called themselves 'miners' regardless of whether they worked overground or underground). 'He likes to display it to encourage people, to show them what they can find if they try hard enough. But no one ever finds as much as Dave Mack.'

There wasn't very much mining equipment in the New 49ers store – it was primarily an office administering the memberships of those who worked the company's claims – but there was enough for me. First, Terry handed me the most important piece of equipment in any miner's kit bag: a gold pan. This is essentially a bowl about four inches high and anything from six to twenty inches across, which has grooves along one edge. When 'pay dirt' – the sand and soil that the prospector hopes will contain gold – is placed into this and washed away with water a layer at a time, the heaviest deposits, including pieces of gold, will become trapped in these grooves.

Next, he passed me a shovel, a plastic trowel, a sixteen-gallon bucket, a wire mesh 'classifier' for separating fine gold-bearing dirt from big but worthless rocks, a pair of tweezers with which to grasp flakes of gold, a small plastic suction bottle to collect them up and some glass vials in which to store them. The vials were very small.

'You ain't gonna need anything bigger than that,' said Terry, reading the disappointment on my face. He was obviously a glass-half-empty sort of person.

'You couldn't fit a decent-sized nugget in there,' I said. I knew I would be much luckier than he was expecting me to be.

Terry looked at me and coughed. 'I wouldn't worry too much about nuggets,' he said.

On top of the items he chose for me, I gathered up a pair of gloves and a short-handled pickaxe.

'You won't need that,' said Terry.

'I want it,' I said, holding the pick to my chest. It felt reassuringly heavy. 'And I'd like one of those – what about that? I could find gold with that.'

I was pointing at a six-foot-long aluminium sluice box. I had heard about these during the court hearing and from my internet mentor, Nathaniel, so I knew roughly how this one worked. It was essentially a flat tray with high sides into which you channelled water, either with a pump or by laying it down in a river's shallows to take advantage of the current. On the bottom of the tray were metal ridges and pieces of nylon carpet deliberately positioned so that when you fed your pay dirt into the water running through the top end, the heavier stuff – including gold – would be caught in the eddies that these riffles created, falling and becoming trapped in the weave of the carpet. The lighter dirt would simply be washed away. Genius.

'You won't need one of those either,' said Terry. 'You need to learn how to pan first.'

I was crestfallen. My bill came to less than a hundred dollars and I didn't feel that was anywhere near enough.

Terry left the store, telling me not to get into trouble and not to buy any more equipment. As soon as he had gone, I went up to the counter and asked Montine if I could meet Dave Mack. I had made a point the night before of asking my fellow miners in Happy Camp about him and, each time I did, something strange happened: they breathed in deeply and gasped, rocking on their heels as if suddenly caressed by a tropical breeze.

'Dave Mack. He's . . . *amazing*,' they would say, glassy-eyed.

'What a guy.'

'He's a genius.'

'A man amongst men.'

'Don't know how he does it.'

'Crazy guy, crazy gold.'

Montine left and returned some minutes later with a man who I guessed was Dave Mack. Everyone wanted to shake his hand and he appeared to be going out of his way to ignore me. It was as if the pontiff had arrived at a football match between the priests and the nuns and everyone had suddenly stopped playing.

'Hello, hi there, hell-ohoo,' I kept saying, but he turned this way and that until I felt about the size of a pea. I followed him ineffectually for several minutes and experienced a feeling of lurking in plain sight, unseen. The former British Shadow Home Secretary Gerald Kaufman once accused me in a newspaper column of lurking in a bush and leaping out to catch him unawares, and I felt I was doing something like that now, only without the bush (about which Kaufman had told the truth).

Eventually, Dave Mack moved into the office and sat on a sofa, unexpectedly turning the glow of his attention on me.

'What can I do for you?' he said, smiling. He was huge and barrel-chested with a close-cropped haircut and sharply defined muscles – his pecs, legs and biceps had weaves in them like subcutaneous plaits of steel. And he was handsome. He was the sort of man with whom other men's wives would like to have an affair. He had a rugged, square jaw and, much later, I found it hard to believe him when he told me he was sixty.

Before I'd had a chance to answer his question, he said, 'I'm an adventure junky. I throw myself into dangerous situations where other people would never dare to go – where they might die.'

I blinked. After a moment's silence I said, 'Can I come out prospecting with you and Rich?' Dave Mack thrust his head back and laughed.

'The last person who came out with me drowned in three minutes,' he said. 'And he was a scuba instructor.' I found myself grinding my teeth. 'Where we're going, the water's so fast that you can't see through it. It's white and we're diving under it with hundreds of pounds of lead attached to our bodies so we don't get swept away.'

'Okay,' I said. 'How about I come out with you for *two* minutes? I've come an awfully long way. I'd love to see some *extreme prospecting*. I promise I won't drown. I promise I won't go anywhere near the water.'

I knew I couldn't keep the last promise – the gold was in the water; I would have to be near the water to watch these guys finding it. But I meant it when I said I wouldn't drown. What kind of suicidal maniac jumps into fast-flowing water with hundreds of pounds of lead attached to their body so that they sink to the bottom of a river while being pummelled by thousands of tons of barrelling spume?

Dave Mack closed his eyes thoughtfully. 'You got water?' he asked after a few moments.

'Yes.' I stifled an urge to say 'Sir'.

'You got food?'

'Yes.'

'You got a hat?'

'Yes.'

'Okay, you can come. But do exactly as you're told and don't touch anything.'

'Yes, Sir!' I said.

5

Not the Gold Rush

Although James Marshall found his gold at the beginning of 1848, the Gold Rush proper did not begin until the following year. Newspaper accounts of strange and wonderful happenings in California appeared throughout spring, summer and autumn of 1848 in small towns all over the United States and the wider world but they sounded too good to be true.

The first report, which went largely unnoticed by the rest of the country, appeared in the *Californian* on 15 March. Located on page two, it read: 'Gold Mine Found. In the newly made raceway of the saw mill recently erected by Captain Sutter on the American fork, gold has been found in considerable quantities. One person brought thirty dollars' worth to New Helvetia, gathered there in short time. California, no doubt, is rich in mineral wealth, [a] great chance here for scientific capitalists.'

On 1 June, the *New York Herald*'s correspondent, going by the byline 'Paisano', wrote: 'We still live and have our being in this "Farest West", with only one serious apprehension, that we are in danger of having more gold than food, for he that can wield a spade and shake a dish can fill his pockets *a su gusto* [as he wishes].'

Gradually, the tone of these reports grew more and more insistent, leaving much of the population of the United States (and, by autumn, the rest of the world) in a state of uncertain excitement – were the stories true or were these reports of gold, of a new El Dorado, the product of someone's fevered imagination?

My friends William Swain and Sarah Royce had both been captivated in 1848 by the earliest accounts of gold finds. (It would be two more years before my third companion, the artist John Borthwick, began to take notice.) William had only recently become a father, to Eliza, while Sarah's daughter, Mary, was at the time aged two. In common with William, Sarah's husband, Josiah, was a farmer. During the summer of 1848 both men wrestled with thoughts of responsibility; neither was the sort of man to expose his family to hardship based on nothing more than rumour – Josiah by taking his to the gold-fields, William by leaving his behind.

On 21 September the *New York Herald* ran a story that read: 'All Washington is in a ferment with the news of the immense bed of gold which, it is said, has been discovered in California. Nothing else is talked about. Democrats, Whigs, free soil men, hunkers, barnburners, abolitionists, all, all are engrossed by the wonderful intelligence. The real El Dorado has at length been discovered, and hereafter let not cynics doubt that such a place exists.'

But did it?

Without doubt, hundreds, then thousands did make their way to the foothills of the Sierra Nevada at the first mention of the word 'gold', but others, more cautious in their outlook, needed more than rumour and hyperbole to spur them into action. Even with rudimentary tools, placer gold was being found in significant quantities within months of Marshall's find. By the autumn of 1848, so much had flooded the lower reaches of the Sacramento Valley that its value had fallen temporarily from fifteen or sixteen dollars an ounce to eleven or twelve, but a full-blown 'rush' was by no means under way. The sober, careful and reticent watching from afar wanted official confirmation before they would risk all on a potentially foolhardy and dangerous journey to a largely unknown land.

What William, Sarah and hundreds of thousands like them needed was formal acknowledgement from their government of the presence of gold. And they finally received it thanks to Lieutenant William Tecumseh Sherman, a West Point graduate who would later go on to distinguish himself in the American Civil War (and, later still, have a tank named after him), and Colonel Richard Mason, the military governor of California.

Sherman had travelled to California in the hope of seeing some action in the war with Mexico, but his posting had been rather boring and administrative. When he was appointed assistant to Mason in Monterey, he expected his days to be only marginally more interesting. However, one day in the spring of 1848, two Americans turned up at his office and demanded to see the governor personally. They said they had been sent by John Sutter. Sherman showed them in and, after a time, he was summoned by Mason: 'I went in and my attention was directed to a series of papers unfolded on his table in which lay about half an ounce of placer gold,' Sherman later wrote. 'Mason said to me, "What's that?" I touched it and examined one or two of the larger pieces and asked, "Is it gold?" I took a piece in my teeth and the metallic lustre was perfect. I then called to the clerk . . . to bring an axe and hatchet from the backyard. When these were brought, I took the largest piece and beat it out flat, and beyond doubt it was metal, and pure metal.'

Sutter had sent these men with a letter requesting the rights to the land on which his saw mill and its mill race were situated. Under the treaty that ended the Mexican-American War, the United States was obliged to recognize the existing property rights of the Spanish-speaking Californios and the Native American villagers. Sutter felt that the land that his saw mill occupied fell into neither of these categories, but he wanted ownership of it to be formalized. Mason said it was far too premature for the rights to any newly acquired Californian land to be determined by the US government. He signed the letter by way of acknowledgement, but sent it back with a general assurance that Sutter should be patient – after all, there were precious few people in Coloma to give him an argument over land rights.

Mason's assurances were made without the benefit of foresight. He could never have anticipated the outrageous behaviour of Sam Brannan, and his classic piece of marketing. You may recall it was Brannan who ran through the streets of San Francisco shouting 'Gold! Gold! Gold from the American River!' in order to generate business for his new store. Contrary to Mason's assurances, Sutter's land was, indeed, overrun as the populations of Los Angeles, San Diego, Monterey and San Francisco, spurred on by word of Brannan's exploits, emptied into the goldfields. He could not repel them, because his men had joined the Gold Rush too.

Thomas Larkin, vice consul in Monterey, was a regular correspondent with Mason. He began to report astonishing gold discoveries as early as 26 May 1848, when he wrote from San Jose, fifty miles south of San Francisco: 'We can hear nothing but gold, gold, gold. An ounce a day, two or three. Last night several of the most respectable American residents of this town arrived home from a visit to the gold regions. Next week they will go with their families, and I think nine-tenths of the foreign storekeepers, mechanics or day labourers of this town and perhaps San Francisco will leave for the Sacramento [River].'

Several days later Larkin, by now in San Francisco, wrote to the Secretary of State, James Buchanan, in Washington: 'I have to report to the State Department one of the most astonishing excitements and state of affairs now existing in this country that perhaps has ever been brought to the notice of the Government. On the American Fork of the Sacramento and Feather River . . . there has been within the present year discovered a placer, a vast tract of land containing gold in small particles.'

Larkin said he understood that twenty thousand dollars' worth of such gold had already been exchanged for tools and provisions – several pounds of which he had seen himself. 'Common spaces and shovels one month ago worth one dollar will now bring ten dollars at the gold regions,' he wrote. 'I am informed that fifty dollars has been offered for one' (the equivalent of about fifteen hundred dollars today). 'Should this gold continue as represented, this town and others would be de-populated. Clerks' wages have risen from six hundred dollars to one thousand per annum.'

A month later, Larkin wrote to Buchanan again, this time with a prophetic hint of panic in his words. He had travelled to the goldfields himself and witnessed miners steadily digging out more than sixty dollars' worth of gold a day. 'If our countrymen in California as clerks, mechanics and workmen will forsake employment at from two to six dollars per day [when the average elsewhere was about one dollar], how many more of the same class in the Atlantic states earning much less will leave for this country under such prospects?' (Of course, it would be many months before these letters arrived in Washington.)

It was the exploits of these first prospectors that were slowly finding their way into newspapers throughout the Midwest, to farming families such as Sarah's and William's, along the Atlantic seaboard and across the globe. They were reaching the ears of Sherman, too.

'I, of course, could not escape the infection, and at last convinced Colonel Mason that it was our duty to go up and see with our own eyes, that we might report the truth to our Government,' Sherman recalled. In June 1848, the pair travelled to San Francisco, then upriver to Sacramento and overland to Sutter's Fort. Describing Sutter and his imperious surroundings, Sherman wrote: 'His personal appearance is striking, about forty or fifty years of age, slightly bald, about five feet six inches in height, open, frank face, and strongly foreign in his manner, appearance and address. He speaks many languages fluently, including that of all Indians and has more control over the tribes of the Sacramento than any man living.'

Of Sutter's outpost, Sherman wrote: 'The fort itself was one of adobe [clay and straw] walls, about twenty feet high, rectangular in form, with two-storey block-houses at diagonal corners. The entrance was by a large gate, open by day and closed at night, with two iron ship's guns near at hand. Inside there was a large house, with a good shingle roof, used as a storehouse, and all round the walls were ranged rooms, the fort-wall being the outer wall of the house . . . Sutter was monarch of all he surveyed, and had authority to inflict punishment even unto death, a power he did not fail to use.'

I could not help but find parallels here with Joseph Conrad's frighteningly all-powerful Kurtz in *Heart of Darkness* (1899). Some visitors to Sutter's compound reported a kind of pastoral paradise; all seemed calm and orderly. He routinely welcomed travellers and provided food and shelter, asking nothing in return, and so his reputation before gold was discovered was as a kind and benevolent man. There are other accounts, however, of him meting out cruel punishments to Native Americans who refused to bend to his will. Some of these say he used local tribespeople as slaves.

Sherman and Mason continued their journey up the American River to a spot that had become known as Mormon

Island, or Mormon Diggings, because it was being worked by three hundred miners led by Sam Brannan who, in a characteristically opportunistic act of vertical integration, had decided to branch out into gold extraction. After Coloma, this was the site of the second gold find of the nascent rush. You will not be surprised to discover that Brannan was first to claim these diggings and was taxing his brethren one-tenth of everything they found.

Sherman and Mason's journey lasted from 12 June to 17 July. By its end they were fully convinced that the United States had acquired gold in abundance with its acquisition of California. Returning to Monterey, on 17 August Mason wrote a report to Brigadier General Roger Jones, the US adjutant general, in Washington, assuring him that everything he was hearing about Californian gold was true.

He described meeting James Marshall, who told him exactly how and where he first found gold, and interviewing miners who told him about fantastic – yet not untypical – finds. Some groups were employing Native Americans and paying them in a variety of 'merchandise' in return for their labour. Mason reported finds in the tens of thousands of dollars. One man showed him the proceeds of just one week's work – fourteen pounds of gold.

'A small gutter, not more than one hundred yards long by four feet wide, and two or three deep, was pointed out to me as the one where two men had a short time before obtained seventeen thousand dollars' worth of gold,' reported Mason. That would be worth about half a million at today's values.

This sudden easy wealth – found so far in the American, Sacramento, San Joaquin, Yuba, Feather and Bear rivers – was resulting in seemingly unsustainable increases in the cost of basic foods. Mason reported that flour sold by John Sutter was expected to rise to fifty dollars a barrel in the coming months, the equivalent today of fifteen hundred dollars. And this was before the Gold Rush had properly begun.

On 18 September, the *New York Herald*'s Paisano wrote:

The whole mass of foreign population struck, not for higher wages, but for none at all – spades and shovels rose from two dollars to ten; tin pans and cups to unheard-of prices; a few considerate turners and blacksmiths remained to make spades and picks, and turn wooden dishes to wash out the sand. These few are now making twenty to forty dollars a day.

The result, in a few words, is that more than half, I think three-fourths of the houses in some towns are vacated. A passage on the launch rose from nothing to four to eight to sixteen dollars. Everyone brought more astonishing news of this El Dorado of rivers whose bottoms were gold, only requiring [one] to step in, scoop up a handful of black sand, move the hand a few minutes in the water, and there remained the pure thing itself.

Near two hundred houses in the town of San Francisco are closed by the owners. Benicia, a small town of a year's growth, had but two men left who were earning thirty dollars a day by the ferry. Monterey is now showing strong symptoms of the gold fever . . . Onward goes this fever, raging strongly in the brains of all, de-populating towns, carrying off men, women and children. A six year old child can gather two or three dollars a day; a man ten to thirty; old and young ladies in proportion.

Paisano concludes: 'No More. You will not believe a quarter of what I have told you, and your readers not a half; the writer is bound to believe much of it – all of it. How long the banks of our rivers will produce gold dust is our affair. Those who have travelled these splendid regions say there is no end to their riches.'

Mason was able to confirm the correspondent's claims.

'The discovery of these vast deposits of gold has entirely changed the character of Upper California,' he wrote:

> Its people before engaged in cultivating their small patches of ground, and guarding their herds of cattle and horses, have all gone to the mines, or are on their way thither. Labourers of every trade have left their work-benches, and tradesmen their shops; sailors desert their ships as fast as they arrive on the coast; and several vessels have gone to sea with hardly enough hands to spread a sail. Two or three are now at anchor in San Francisco, with no crew on board. Many desertions, too, have taken place from the garrisons within the influence of these mines; twenty-six soldiers have deserted from the post of Sonoma, twenty-four from that of San Francisco, and twenty-four from Monterey.
>
> I have no hesitation now in saying, that there is more gold in the country drained by the Sacramento and San Joaquin Rivers than will pay the cost of the present war with Mexico a hundred times over. No capital is required to obtain this gold, as the labouring man wants nothing but his pick and shovel and tin pan, with which to dig and wash the gravel, and many frequently pick gold out of the crevices of rocks with their knives, in pieces of from one to six ounces.

Mason sent with his report a tin containing 230 ounces of gold, couriered by one Lieutenant Lucian Loeser, just to prove his point. Loeser, who must have been an uncommonly honest man, set off with all of this gold in August 1848 and did not arrive in Washington until late November. After considering Mason's report for several days, President James K. Polk decided to make its findings public.

On 5 December, more than ten and a half months after

Marshall's discovery, President Polk told Congress: 'The accounts of the abundance of gold in [California] are of such an extraordinary character as would scarcely command belief were they not corroborated by the authentic reports of officers in the public service who have visited the mineral district and derived the facts which they detail from personal observation.'

The gold sent by Mason and Sherman was put on public display while the newspapers went wild. 'El Dorado of the old Spaniards is discovered at last,' reported the *Herald*. 'We now have the highest official authority for believing in the discovery of vast gold mines in California, and that the discovery is the greatest and most startling, not to say miraculous, that the history of the last five centuries can produce.'

Californian gold had the presidential seal of approval. The brakes were off and would-be Argonauts the length and breadth of the country – then the world – began laying plans to travel to California, William Swain and Sarah Royce among them.

El Dorado was real after all. To strike it rich, all you had to do was to get there.

6

Extreme Prospecting

Dave Mack was a former US Navy SEAL, the toughest of the tough, but in his book *Extreme Prospector*, on whose cover he can be seen holding up a gold nugget and revealing an immaculately shaved armpit, he admits to a softer side that made me warm to him.

As a child, he was brutally disciplined by his father, a nuclear submarine commander who would drink too much and return from missions to the family home in Waterford, Connecticut, principally to beat Dave, his mother, two brothers and sister. The boy McCracken would often go to school with a black eye and swollen lip from his father. He was shy and lacked confidence, spending too much time alone and, without direction, he would get into trouble – nothing serious, just pilfering construction sites 'to make forts out in the woods.'

McCracken Snr sapped the boy's confidence to the point where he felt worthless. Father repeatedly told son that he was no good and would amount to nothing. But the boy did amount to something, thanks to a deep love of water. First, this led the young McCracken down to the sea to build a small lobster-fishing business; then it took him into the office of the US Navy recruitment service. He was allowed to take a crack at becoming a SEAL (so-called because they operate on sea, air and land), and, to the astonishment – possibly chagrin – of his father, he made it through the most difficult military training course in the world.

Very few people complete the SEAL Basic Underwater Demolition, or BUDS, training course; Dave Mack's class was reduced from fifty-eight candidates to just eight within three weeks. It was brutal, punishing and aimed at drowning, crushing or freezing you to death. You could back out at any time, but once you did, you were history. He graduated and later served during the evacuation of American personnel from Saigon at the chaotic denouement of the Vietnam War. I came to the conclusion that he was a man of considerable moral and physical fibre.

When Dave Mack was discharged from the service in the late 1970s he began to direct the skills of endurance, problem-solving and determination, which he had developed as a SEAL, towards finding gold. Adventures followed in California, Canada, Venezuela, Cambodia, Borneo, Thailand, Madagascar, India and the Philippines, during expeditions that would see him scoop up more gold than probably any other small-scale miner alive.

Being allowed into his world after meeting him for just five minutes was a privilege I perhaps did not fully appreciate at the time. Rich Krimm arrived at the New 49ers HQ and seemed floored after being told that I was prospecting with them that day. He stared at Dave Mack with something approaching amazement and a look that said, 'Are you absolutely sure this is a good idea?'

I smiled at Rich. I had a feeling that he wanted to strand me with the wild miners at Independence Camp and leave me there, out of the way, while he and Dave Mack blazed their own secretive, golden, trail. Before he could voice any objections, I said, 'I won't get in the way.'

I went out into burning sunshine – it was 109 degrees – and climbed into my four-by-four. Rich had told me that he and Dave Mack would emerge shortly from the back of the New 49ers building but I was convinced they would try to give me the slip. There was gold at stake and I couldn't imagine a situation involving it that would play out simply and without deviousness. I put on my sunglasses and waited.

All I knew was that we would be going downriver, probably to some isolated spot far from any other human presence. The night before, around my makeshift table, I had asked Terry, Craig and Tom what I needed to know about the wilderness. What did I need to look out for? Bears, I was thinking; I had been confused for some time about how best to handle bears. Should you stand up and make yourself big when confronted by a grizzly? No, that was what you did with a brown bear. Or maybe that was a black bear. You should run downhill – they can't run downhill because of their short hind legs. Or was it uphill? Yes, that was what you did with a black bear. Or you should climb a tree – at least one species of bear can't climb trees, but which one was it? And with a brown bear you lie down and pretend to be dead until it loses interest. Or was that what you did with a grizzly? No, you jumped into the river and swam – they're not very good swimmers. Or were they excellent swimmers? I couldn't remember. The truth was, I had no idea – and the thought of confronting something with claws and fangs that weighed considerably more than a tumble dryer filled me with terror.

The three men looked at one other.

'If you see a bear? Best thing is . . . maybe wave and take a picture,' said Terry. They laughed. I maintained the utmost seriousness.

'Honestly,' said Craig, 'you probably won't see one and if

you do you'll be lucky. It will most likely be on the other side of the river and won't be interested in you. They don't eat people. They hide from people. Unless it's a mother with cubs. If you get anywhere near those cubs, or between the mother and the cubs, she'll rip you to pieces.'

There were no grizzlies or brown bears in the area, only black bears and these were not particularly aggressive. Just twelve black bear attacks had been recorded by the California Department of Fish and Wildlife since 1980, but since a male could weigh up to 550 pounds and was easily capable of carrying off a deer, I had no desire to become unlucky number thirteen.

No one had died during these attacks, but people had died in other attacks elsewhere. In May 1978 three teenagers were killed by a black bear in Algonquin Provincial Park in Ontario, Canada, while enjoying a day out fishing. In 1997 in the Liard River Hot Springs Provincial Park, again in Canada, a black bear killed a mother and the man who had intervened to try to protect her and her child.

As a city slicker from a country where the worst thing that could happen in the wild was a bee sting, I found it impossible to be casually unafraid of these creatures. The official list of attacks was full of inconsistencies and incidents that did not involve cubs, and seemed to be random and vicious. Of these, the most worrying from my point of view occurred in 1993 in San Bernardino County when, in two almost identical attacks, a particularly hungry or delinquent bear crept up on dozing teenagers, took their respective heads in its jaws and dragged them from their sleeping bags before eventually dropping them, probably in need of a change of trousers and with a life-long aversion to the outdoors. I looked at my flimsy tent.

The oddest of these reports involved a man who was bitten in Mono County in April 1996. The official version reads simply: 'A man received a bite on the buttocks from a young bear. Further details are not known.'

I tried to imagine how this incident made it, so incompletely, into the log. In the version that played out in my mind, a high-powered lawyer was hiking through the forest and was gnawed, playfully, on the backside before limping, his expensive walking trousers bearing four puncture marks, into the local Department of Fish and Wildlife office.

'I've been bitten on the ass by a bear,' he says. 'Does that constitute an attack?'

'Do you feel as if you've been attacked?' asks the ranger.

'I feel like I've been bitten on the ass.'

'You were lucky – could've gone for your throat.'

'It was only this big.' (Bends down with flat of palm at knee height.)

'So you were bitten on the ass by a two-foot bear cub?'

'Yeah.'

(Picks up pen, sighs.) 'Name?'

'Doesn't matter.' (Shuffles out into wilderness.)

While I waited for Dave Mack and Rich, I revisited all the other advice that Tom, Terry and Craig had given me on the subject of dangers.

'You're more likely to be attacked by a mountain lion than a bear,' Tom had said. 'That's still pretty unlikely but they have been known to kill people.'

'Nah,' said Terry. 'It's rattlesnakes you should be worried about. Get a good stick and hit the ground in front of you. Don't want to take one o' those guys by surprise. If he hears you coming, he'll get his tail up and start rattling and you can back off slowly. You get bitten out there and don't get help, you could die.'

The others had nodded sagely for a few moments before Craig broke the silence.

'Well, that's true, but it's poison oak that's really gonna get you,' he said. 'Poison oak's everywhere and it itches like a bitch if you touch it. Sometimes blisters too. It won't kill you but, boy, will it ruin your week. You know what poison oak looks like?'

I shook my head.

'It has three leaves like this,' he said, spreading out the middle three fingers of his right hand. 'You better learn to spot poison oak.'

For a few moments, while I added this to my list of fears, all I could hear was the sound of cicadas and my unsteady breathing.

'Dehydration,' said Terry, suddenly. Tom and Craig looked up and nodded. 'People get carried away in the heat. Work too hard. You're losing fluid all the time and if you don't replace it, you pass out. If there's no-one to help you, you're a goner.'

I looked at my feet and took a deep breath. Water, I thought. Always water. Never be without water and don't take any chances. And salt. Don't forget salt. You'll need that. And a stick. And maybe a gun.

'Ticks,' said Tom suddenly, with a shiver. Everyone turned in his direction. 'That's what scares me most. They hang on trees and wait for you to pass by. Then they drop on you and burrow under your skin, sucking your blood. If you don't spot them on your body, before you know it you've got Lyme disease and that can go to your joints, your heart, your brain.'

Craig and Terry nodded, pulling faces. I swallowed hard.

'Tarantulas . . .' began Craig.

I was staring up at the mountains, wondering whether I was ready to head into the wilderness with Rich and Dave Mack when I was shaken by the honk of a horn. They weren't trying to give me the slip; they were slowing down and gesturing me to follow them. For a fraction of a second, as they overtook and led me to Highway 96, our cars were parallel and before I was blinded by the sun I saw the silhouette of two men with their mouths open as if yawning or laughing. I couldn't be sure which.

I followed the extreme prospectors south-west along the Klamath, the river's clear waters rolling first deep and calm and slow, then suddenly wild and white, roaring with deadly power. Young willow trees drenched in spray hugged the river-bank under rainbows, while above and in the distance, swollen forests of fir and pine floated on puddles of hazy air. We were in the Klamath-Siskiyou Wilderness, a region of unspoilt beauty, little troubled by roads, which covered an area of almost ten million acres in south-western Oregon and north-western California. It was the largest stretch of open, virtually untouched country on the Pacific coast of the United States.

I had anticipated a long and spectacular drive, but the men pulled off the road after just twenty miles and steered underneath a huge green bridge, elephantine in its inappropriateness, which straddled the river and carried a road that seemed to go nowhere beyond the other side.

'Welcome to Independence Camp,' said Rich, slamming his car door.

I looked up. So, they were leaving me here after all, I thought. I felt crushed. Rich must have worked on Dave Mack and convinced him that I would be a liability. I kicked the ground and wondered how I could convince them to take me along.

The camp was almost invisible from the road and con-sisted of a handful of old-fashioned canvas tents and several

RVs on a small dusty plateau with lovely views of the river below. Our arrival attracted the attention of the few inhabitants who weren't already out prospecting. The ones I noticed first were three American pit bull terriers, two fully grown and one pup. Almost immediately, smiling under a camouflage hat, shirtless, tattooed and with a bandanna round his neck, came the dogs' owner. He had a roll-up cigarette protruding from a luxurious white goatee beard and was muscled, wiry and intimidating.

'Hi guys,' he said, and shook hands with Rich. His name was Duane. Dave Mack nodded and began unhooking the aluminium boat they had been hauling. I had the impression that McCracken never felt the need to make small talk with anyone. The dogs circled my boots, sniffing, and I gave them the back of my hand unthreateningly until they grew bored and wandered off. Rich introduced me to Duane and he took a look at me, taking the smile off his face as he did so.

'Where's Wild Bill?' asked Rich.

'Oh, he's out there somewhere,' said Duane, spitting on the ground. 'Says he found a good pay streak and wants to make the most of it.' (Normally I would have made inquiries as to the character of an individual called Wild Bill, but right now it somehow didn't seem appropriate.)

Rich nodded and I felt the two men would never have spoken to one another outside the world of gold prospecting. Rich was suave and elegant and his clothes – Gant, perhaps, or Abercrombie & Fitch – were pressed and neat; Duane was every inch the feral miner I had allowed myself to imagine.

'How're your ribs?' asked Rich.

'Still cracked but coming along fine. Be back in the saddle in a coupla days.'

Rich turned to me. 'Duane broke some ribs when a trolley full of equipment fell on him as he was hoisting it down to the river. Would've killed most guys.'

I smiled nervously and made my way to the riverbank, where Dave Mack was winching the boat into the water.

'Is this it?' I asked him. Although in 'wilderness', we had not travelled far and our location did not feel particularly wild.

'We're going a few miles downstream,' he said, pointing into a deep canyon that wound farther away from the road. I wondered if I would be going too.

'About fifteen years ago, some buddies of mine pulled about a thousand ounces out of a spot down there – worth about two hundred thousand at the time – and no one's been back there since, so this summer we figured we'd take look. We've been there a while but we haven't found anything. Trying a new area down there today and if we still find nothing we'll move on.'

The boat contained the men's wetsuits, weights and some air lines, tools, fuel and steel boxes that I imagined housed guns and dynamite, pumps and winches, but I really didn't know. My heart missed a beat when Rich told me to get in. I asked where to sit so I'd be out of the way and decided to shut up while the partners prepared the boat. It wasn't until Dave Mack put the motor in reverse and we started backing away from the bank that I began to believe I wasn't being left behind.

We quickly gathered speed and I had to hold on to my straw hat while Dave Mack steered us through shallows and over gentle rapids that must have been caused by rocks but which our boat's narrow draft cleared with inches to spare. Rich had noticed my unease and was doing his best now to make me feel more comfortable.

'Dave's been running this portion of the Klamath for more than thirty years,' he said. 'He knows every rock in the river.'

Even Dave Mack seemed to be loosening up, smiling as I took his photograph. The west bank of the river was almost vertical to the road about fifty feet above, while on the gentler eastern side the scrub climbed slowly and undisturbed to the treeline. I looked up and could have sworn I saw an osprey wheeling across an otherwise empty blue sky. Must be the heat, I thought.

After only a few minutes we reached the site where the prospectors had moored their dredge the day before at the close of a fruitless day. Dave Mack spun the boat through 180 degrees and sped into the rapids towards the floating platform. It was a grey and green-coloured composite plastic raft at the rear of which was a pump connected to two hoses about four inches in diameter. At the front were two more pumps; between them, they would power two vacuum hoses, two air lines and a black box with a filter that rocked back and forth to separate heavy dirt from lighter soil. They had been using this piece of equipment at several sites on the inside of a bend in the river, but without success.

Choosing the inside of the river bend was a classic prospecting technique. Water travels slower on the inside than on the outside (imagine a record spinning round; the speed is lower at the centre than at the edge), so during a raging flood this is where it will lose the kinetic energy that gives it the power to carry gold from seams high up in the mountains, and where the metal will fall to the riverbed.

'It goes against all the rules, but we're moving away from the inside bend because we haven't been able to find any rich gold deposits over there,' said Dave Mack. 'The prospectors that were working here found a lot of gold. Maybe they cleaned it out but I can't believe they didn't leave something behind in the real fast water where we're looking.'

The men were hoping that the earlier prospectors wouldn't have fancied facing the dangerous middle or outside of the river, and the day before their hopes had been partially rewarded when they dived down and found several huge boulders that would have halted the passage of gold washed down during floods of yesteryear. The plan was to look underneath the boulders. Of course, prospecting on the outside would mean diving into even faster currents, into truly violent white water. I thought they were insane.

They put me ashore and towed the dredge as close as they could to the rapids before throwing me a line with instructions to tie it securely to the base of a young willow. The bank was rocky and there were no mature trees and no flat ground, only boulders, saplings and grasses that hid anything resembling open space. I had to hop from boulder to boulder, tugging the line behind me through a habitat that was beyond perfect for rattlesnakes. Meanwhile, on the platform, Rich and Dave climbed into their wetsuits, fired up the compressors and began to put sixty-pound lead weight belts around their waists. Then they pulled down their face masks, bit on their mouthpieces, slipped into the water and disappeared.

Suddenly, I was completely alone except for the thrum of the pumps and the hydraulic slurping of the box filter rocking back and forth like a rusty iron lung. I looked around and came to the conclusion that this would be the perfect location for a bear attack. I picked up a stick and began tap-tapping between the boulders on which I stood, wondering if I would hear the sound of a rattler's tail above the din of the pumps. Twenty feet from the river, the bank rose suddenly into a fifty-foot bluff that was too wild and steep to climb. There was no escape.

Before lowering himself into the rapids, Dave Mack had told me he wasn't kidding when he said someone had died after just three minutes in the water with him. 'He did die, but Rich saved him,' he said, patting Rich on the back.

'Yeah,' said Rich. 'I did CPR on him for about fifteen minutes before he started breathing on his own, then we just stayed with him until the paramedics came. But I'd say he was probably dead for a while.'

The story was ringing in my ears as I tried to make out the figures of the men in the water. They were just below me – or within a few yards of where I was standing – but all I could see was some vague movement where the river slowed or where the divers strayed from the white water. They were carrying tools, were weighed down by lead belts and over their shoulders they hauled vacuum hoses through which they sucked sediment from the cracks into which they reasoned gold could have fallen. And all the while they were inching themselves forwards by their fingertips, from rock to rock on the riverbed and against the raging force of the river. After about twenty minutes, the pair surfaced near the bank, but only to take on more weights, and, ignoring me, they dived back to the bottom, ten feet or so below.

There was no shade on the bank and the temperature soared while I visited and revisited the bottled water in my rucksack. Nor was there anywhere comfortable to sit, so I found myself hopping from rock to rock to alleviate my sense of boredom as the minutes turned into hours. My chief pastime was to check and recheck the knot I had tied in the line that held the mining platform in place. If it came loose and the platform was suddenly wrenched into the current, the men's airlines would be tugged away and they would drown if they failed to shed their weight belts fast enough. If they did break free, they would be at the mercy of the white water as they tried to surface. Either way, it was more responsibility than I needed.

I wondered whether my role, securing the all-important line to the floating platform and watching for bears and rattlesnakes, might entitle me to a share of any of the gold that the prospectors found – but it seemed academic when, after two and a half hours in the water, they hauled themselves into the shallows and emerged with long faces.

'Should've been something there – you know where I mean?' Dave Mack asked Rich, gasping.

'Yes, I do. That big boulder? I know the one. I saw it too but there was nothing.' Rich looked exhausted, his chest rising and falling as if weighed down by something invisible but heavy.

'I thought we'd get colour in the crevices on the downstream side of that boulder. I dug underneath it but there was nothing there.'

'Colour' is the prospectors' term for gold.

'You think it's cleaned out?'

'No. No way anyone's been under there before us. Nobody but us crazy enough to go under that water. Maybe there's just nothing there. But, man, there should be.'

I jumped from a boulder on to the platform, my rucksack on my back and the men passed me masks and belts and tools as they lightened themselves in the water, but I said nothing. Their moods were dark and they barely acknowledged my presence. I did not want to be seen as the cause of this bad luck. I did not want to be their Jonah.

They climbed on to the dredge and Dave Mack pulled the boat alongside. We all jumped in and Rich and I began rummaging in our bags for the sandwiches we had brought while Dave Mack stood at the prow, staring into the river as if looking for something he had lost.

'I didn't expect *nothing*,' he said after a while. Rich looked at me.

'They call him "Ounce a Day Dave" but he's had terrible luck all summer,' he said. I thought back to Terry's lazy prospector but there was no irony in McCracken's nickname – he'd had a 'bad' summer, but this season he had still found all the gold I saw in the pot at the New 49ers HQ.

'We haven't found anything for weeks and it's getting to him. Hell, it's getting to me too. We knew this might be a long shot, but with those boulders and the speed of the river, we should be seeing something down there.'

I imagined the effort of staying underwater for more than two hours against a current this fast, moving rocks, sucking up sediment with a heavy hose on your shoulder, prising boulders from the riverbed with a crow bar and toting a hundred pounds of lead across your back. No wonder they were in such good shape.

'They say you make your own luck in life, but it's only partly true with gold,' said Rich. 'You can work as hard as you like, but if the gold isn't there, no amount of work will make you lucky.'

I thought this would be the cue for us to pack up and go home, but almost in unison the prospectors stood up, pulled up their wetsuits, put on their masks and belts and jumped in again, vanishing into the wild, wild water. The speed of their departure took me by surprise. There must have been some signal that I had missed, and that made my sense of isolation above water even greater. I repeatedly scanned upriver and down, wondering whether exposure to such beauty could, with repetition, fail to take one's breath away. Could you become inured to it? I looked up and saw the osprey again – I was later to find that there was a pair nesting nearby – and a kestrel carrying a struggling mouse in its talons. If you were born here and this was your default location, would you even know it was so uncommonly beautiful?

I found a flat slab of granite in front of a grandstand of blackberries and lay down, pulling my hat over my face, but I couldn't fall asleep. Each time I closed my eyes I imagined a bear creeping up on me. The sound of the rocker groaning rhythmically occupied every thought like a hypnotically high-pitched voice.

I expected another wait of two hours or more, but after just fifty minutes the divers burst to the surface. I leapt up from my rock in the expectation that they had decided they were wasting their time; instead, they pulled out their respirators and began to whoop.

'Now *that's* what I'm talking about!' yelled Dave Mack.

'Oh man! Did you see that? Did you see that?' shouted Rich.

They were high-fiving and dancing little jigs in the shallow water.

'Pockets, man. I'm talking *pockets* of nuggets! Did you see them?' Dave Mack turned to me. 'Underneath the boulders down there, there are pockets just filled with gold – *filled* with *gold*.' He high-fived Rich again. 'But I gotta tell you, it's dangerous. The water, it wants to rip the mask off your face and the regulator out of your mouth.'

Rich looked up at me, standing on the bank. 'I was getting worried,' he said, spitting out river water. 'Been six weeks since we found anything – anything! And now this. Oh, man! Oh, boy!'

I found myself smiling. I hadn't seen anything yet, but I felt happy for them. The tension in the boat earlier had been uncomfortable. I assumed it was the withdrawal effects of gold fever; if you were used to finding it, then you needed that fix. Now there was laughter as they stowed their respirators and weight belts. They had suctioned up what gold they guessed was in the riverbed sediment and that would be in the processed dirt graded by the rocking box, but there was also a smaller sample tray that they could remove for a superficial check on how successful they had been. They undid that and each scooped up some dirt and dropped it into a pan, filling the pan with water and violently shaking it so that any gold would fall to the bottom, then gently washing away the surface layers into the river over and over again until all that was left were the heavy deposits – iron-rich black sand, and gold.

Within seconds, nuggets, pickers and flakes emerged through the sand, their juxtaposition with the blackness making them blindingly bright, white-yellow under the beating sun.

When you first see gold in a pan, your guts are rearranged and they never right themselves. Your heart presses urgently against the inside of your chest. Your breathing quickens. You wonder how a person could ever be taken in by fool's gold, because there is nothing, nothing you can imagine, like the sight of colour in a pan. For four billion years, the shining object winking at you now has existed, unseen, first as molten metal deep in the earth's mantle, then as the spew of tectonic convulsion, from geothermic spasm, forced, pushed relentlessly, ever-closer to the surface to cool and be locked in petrified quartz veins, hidden for millennia and then undone by glacier and weather and water, until finally washed clear and pulled down, down, down by gravity to the deepest place it could find, here, in this river, under this boulder and into your pan.

Rush by Land, Roll by Sea

Sarah Royce and William Swain set off for California nineteen days apart – William on 11 April 1849 and Sarah on 30 April – with President Polk's words ringing in their ears.

The president's announcement, his confirmation of Californian gold in huge quantities, had reverberated across the United States, then the world, with consequences never before witnessed on such an international scale. On lonely homesteads, in villages, towns, cities and across countries beyond the seas, it was all people talked about. Newspapers were filled with stories of vast gold discoveries – and in the early days some lucky prospectors did, indeed, pick up pounds of the stuff with hardly any effort at all.

There were reports and pamphlets on how to travel to the goldfields, what provisions and equipment to take, which routes were regarded as safe, and which were quicker but riskier. Some of what passed for knowledge was indeed useful; more still, in the early days, was conjecture, for at that time relatively few people had travelled the two thousand miles across plain, desert and mountain to California from the recognized trail-heads along the Missouri River in the Midwest. Since the early 1840s, some intrepid settlers had blazed the Oregon Trail westward but it was still feared as a hard and potentially deadly route that took upwards of six months to complete. Until then it had been regarded as a one-way journey; most of those who took it had no intention of returning. The route into California pushed farther still across deadly deserts and mountains. It was not for the faint-hearted.

The lure of gold transfixed dreamers and planners alike, from the practical and practised to the naive and hapless. Within days of President Polk's address, men began preparing noisily, raiding their savings, mortgaging their homes, grouping together in 'companies' or packing for solitary journeys. For some, great riches were the draw; for others, the possibility of finding enough gold to clear debts on the farm, to provide for a simple but comfortable life, was the stated aim. Of course, if such modest ambitions were quickly realized, would the men who expressed them really walk away from digging just one more ditch?

It was a great adventure and it would lay the foundations for what was to become the American dream. The reality of Manifest Destiny would spread its wings and fly to the West Coast in a matter of months rather than crawl there, as it surely would have, over several decades. The extent to which gold managed to populate the western seaboard and further unite – then divide – the states of America, should not be underestimated.

Within weeks, entire towns and hamlets in the Midwest were left without men. The burden on women and children was sudden and extreme. They were tasked with surviving while their gold-fevered husbands, fathers and brothers rushed headlong into danger. And make no mistake, it was dangerous.

There was the half-year journey to contend with, during which time the average 49er could die from cold, heat, disease, accident, drowning, starvation, dehydration or dispute, either with his fellow prospectors or with the Native American tribes he disturbed along the way. Then there were the living conditions, appalling in the extreme, once he reached the goldfields, and the temptations to which he might succumb: the boredom, the drink, the gambling and the erosion of memory, moral rectitude and the sense of responsibility with which all had set out but which some would cast aside. Many never returned – from choice, circumstance or death.

Sarah had been educated at an academy in Rochester, New York State. She had a studious and inquiring mind and an unshakeable sense of right and wrong that had been drilled into her during a strict Christian upbringing. She also had a Victorian sense of duty. When Josiah announced that they were to leave their home in Iowa, twenty miles west of the Mississippi, to head for California, she seems to have accepted the decision with less trepidation than unquestioning obedience, especially considering she had never before slept in the open.

The couple were accompanied by a male companion and, although they would later join a wagon train with more experienced travellers, at this point their journey was informed only by a government guide to western migration written by a young captain, later major, in the Corps of Topographical Engineers, based on his travels through Oregon and into California in 1842 and 1843. His name was John C. Fremont, an adventurer who came to be known as 'The Pathfinder of the West' for the expeditions he led that opened up these mysterious new territories. He was an explorer who was as nakedly ambitious and ruthless as he was brave, swashbuckling and feckless. Luck, as we shall see later, was drawn to Fremont as surely as it was repelled by John Sutter.

'Our outfit consisted of a covered wagon, well loaded with provisions and such preparations for sleeping, cooking etc., as we had been able to furnish, guided only by the light of "Fremont's Travels", and the suggestions, often conflicting, of the many who, like ourselves, utter strangers to camping life, were setting out for the "Golden Gate",' wrote Sarah many years later, drawing on entries from what she called her Pilgrimage Diary. 'Our wagon was drawn by three yoke of oxen and one yoke of cows. The latter being used in the team only part of the time. Their milk was, of course, to be a valuable part of our subsistence.'

It took no time at all for the realities of the open road to hit Sarah. Having been brought up in towns and cities, it seems not to have fully dawned on her that her new life might be very different. To her credit, she did not complain.

'The [first] day turned out by no means unpleasant,' she recalled. 'Our first noon lunch was eaten by the whole party, seated in the front part of the wagon while the cattle, detached from the wagon but not unyoked, grazed nearby. After a short rest we again moved on. The afternoon wore quietly away, the weather being rather brighter and warmer than the morning – and now night was coming on.'

Suddenly, the reality of what was happening to her seems to have struck home. The enormity of the landscape, and the scale of the journey that lay ahead, must have seemed terrifying.

'No house was within sight,' she wrote. 'Why did I look for one? I knew we were to camp; but surely there would be a few trees or a sheltering hillside against which to place our wagon? No, only the level prairie stretched on each side [of] the way. Nothing indicated a place for us – a cosy nook in which for the night we might be guarded, at least by banks and boughs. I had for months anticipated this hour, yet, not till it came did I realize the blank dreariness of seeing night come on without house or home to shelter us and our baby-girl. And this was to be the same for many weeks, perhaps months. It was a chilling prospect and there was a terrible shrinking from it in my heart; but I kept it all to myself and we were soon busy making things as comfortable as we could for the night.' Where others might have buckled and pleaded to return to the comfort of the family farm, Sarah remained calm and determined, stoic and practical.

As I lay in my own camp, worrying about bears, I wondered what I would have done as a young man in 1849. Would I have had the fortitude to make it to California? Even as they progressed, Sarah had asked herself the same question.

'At first the oppressive sense of homelessness, and an instinct of watchfulness, kept me awake,' she wrote. 'Perhaps it was not to be wondered at in one whose life had so far been spent in city or town, surrounded by the accompaniments of civilization and who was now, for the first time in her life, "camping out." . . . It soon became plain that the hard facts of this pilgrimage would require patience, energy, and courage fully equal to what I had anticipated when I had tried to stretch my imagination to the utmost.'

William Swain was not so unprepared, which was just as well as his journey was about six hundred miles longer if a crow were to fly west to Sarah's home in Iowa from William's farm in Youngstown, New York State. He lived on the western edges of the territory in a cabin built by his father, Isaac, an Englishman whose original home on the site had been razed to the ground by his British countrymen in the War of 1812. His new American countrymen had provided the replacement courtesy of a two-hundred-dollar grant from the state of New York, a sum distributed generously to the 'sufferers on the western frontier'.

After Isaac died in 1838, William farmed there with his brother, George, and mother, Patience. In 1847 he married

Sabrina and the following year she became pregnant. By the time their daughter, Eliza, was born, William was already reading about the great gold discoveries in California and growing restless to be there. After much discussion with Sabrina and George – who would have to be relied on to remain home, run the farm and provide for their mother, Sabrina and Eliza while he was away – it was decided that William should head for the goldfields to find riches enough to keep all his family in comfort for the rest of their days.

Several close and trusted friends from in and around Youngstown agreed to form a small company with William and, after an unexpectedly difficult goodbye, off they set. Describing his departure, William wrote in his diary: 'All my things being ready last night, I rose early and commenced packing them in my trunk, preparatory to leaving home on my long journey, leaving for the first time my home and dear friends with the prospect of absence from them for many months and perhaps for years.

'Among these are an affectionate wife to whom I have been married less than two years and an infant daughter ten months old, to both of whom I am passionately attached; an aged mother who from her great age – seventy-one years – much probability arises of never seeing again on this side of the grave ... and an older brother to whom I am deeply attached ... Being two years older than myself, he has been my adviser and guardian from youth up, at once a father and a brother.'

Leaving behind three generations of family with no guarantee of being reunited was almost too much for William to bear. He was not alone; all over the United States there were similar partings. And many, many families would never be whole again.

'I had fortified my mind by previous reflection to suppress my emotions, as is my custom in all cases where emotion is expected,' wrote William. 'But this morning I learned by experience that I am not master of my feelings in all cases. I parted from my family completely unable to restrain my emotions and left them all bathed in tears.'

There were three ways to reach California: by land, sea or a combination of the two. The sea routes headed north from the western ports of South America, west from Europe, east from China and Australasia, and south from the eastern ports of the United States and around Cape Horn at the tip of South America for the northward journey up the west coast of North America.

Many east coast Argonauts cut two months off the journey by sailing south to the Caribbean sliver of land that is Panama, negotiating the Chagres River to Cruces then travelling overland to Panama City on the Pacific coast before boarding a ship north to California. From New York, this was a journey of just over six thousand miles compared with almost seventeen thousand if they went round Cape Horn. The route was quick and convenient when it worked – when the ship on which one had booked passage actually turned up. When it did not, it involved weeks or months of waiting in Panama City for passage on ships to California in terrible, disease-ridden conditions that led to the deaths of hundreds. The problem, of course, was that many vessels booked in good faith in New York or Boston never arrived; their crews had left them and gone to the goldfields.

This was the route favoured by John Borthwick. A native of Edinburgh, Borthwick was good-humoured and irrepressible.

The fact that he was fit, young and already a seasoned traveller stood him in good stead, not least in Panama City, where good health and the wherewithal to maintain it might save one's life.

. A small town of narrow streets and churches, the city had been suddenly transformed in 1849 by the arrival of many thousands of people with whom it simply could not cope. Argonauts slept in doorways, under blankets in the street. They drank bad water and ate rank food. Disease was rampant, and was it any wonder when travel guides offered such erroneous advice as: 'Bear the heat, bear the mosquitoes, do anything rather than expose yourself to the night air which is the source of every illness in the climate'?

The last things travellers through Panama needed to bear with were mosquitoes; they should have been doing their utmost to avoid them. During the earliest days of the Gold Rush, hundreds died along the isthmus route from cholera, typhoid and mosquito-borne diseases such as yellow fever and malaria. While Borthwick was careful about what he ate and drank, and how he conducted himself, others who had never been farther than the boundaries of their farms were not so, and many paid for their ignorance.

'There was here at this time a great deal of sickness and absolute misery among the Americans,' Borthwick wrote. 'Diarrhoea and fever were the prevalent diseases. The deaths were very numerous, but were frequently either the result of the imprudence of the patient himself or of the total indifference as to his fate on the part of his neighbours, and the consequent want of any care or attendance whatever.

'The heartless selfishness one saw and heard of was truly disgusting. The principle of "every man for himself" was most strictly followed out, and a sick man seemed to be looked upon as a thing to be avoided, as a hindrance to one's own individual progress.'

The Gold Rush inflated Panama City's population of around

eight thousand by as many as three thousand gold-seekers at any one time, principally Americans, awaiting transportation to California, and waiting for it for up to three months. Bored and impatient, they tried to make themselves at home by taking over the running of the place.

'The Americans, though so greatly inferior in numbers to the natives, displayed so much more life and activity, even in doing nothing, that they formed by far the more prominent portion of the population,' wrote Borthwick. By the time he arrived there in 1851, it was less chaotic than in the early days of the Gold Rush, but it remained far from pleasant: 'The main street of the town was densely crowded, day and night, with Americans in bright red flannel shirts, with the universal revolver and bowie knife conspicuously displayed at their backs.

'Most of the principal houses in the town had been converted into hotels, which were kept by Americans and bore, upon large signs, the favourite hotel names of the United States. There were also numbers of large American stores or shops of various descriptions, equally obtruding upon the attention of the public by extent of their English signs while, by a few lines of bad Spanish scrawled on a piece of paper at the side of the door, the poor natives were informed, as a mere matter of courtesy, that they also might enter and buy.'

In common with all the places where gold miners congregated, gambling was rife, as was drinking and whoring, all pastimes designed to alleviate the monotony of waiting for a passage to San Francisco.

Borthwick wrote: 'Life in Panama was pretty hard. The hotels were all crammed full; the accommodation they afforded was somewhat in the same [primitive] style as at Gorgona [along the route from Chagres], and they were consequently not very inviting places. Those who did not live in hotels had sleeping quarters in private houses, and resorted to the restaurants for their meals, which was a much more comfortable

77

mode of life. Ham, beans, chickens, eggs and rice were the principal articles of food. The beef was dreadfully tough, stringy and tasteless, and was hardly ever eaten by the Americans as it was generally found to be very unwholesome.'

After a relatively trouble-free passage, Borthwick arrived in San Francisco in the spring of 1851.

The overland route was divided into two distinct stages. First came the journey to the Missouri River trailheads, to the 'out-fitting towns' of Independence or St Joseph, both in Missouri, or Kanesville (today called Council Bluffs) in Iowa. As the name suggests, these towns were the places where teams spent hard-earned cash buying everything they would need – from oxen and cows to wagons, food and weapons – in preparation for the second stage of the journey into the truly Wild West.

Argonauts travelling there from farther north or east often found it quicker to head by steamer first, north through the Great Lakes of Canada before sailing south down the newly opened Michigan–Illinois Canal and the Illinois River to St Louis, then heading due west through Jefferson City to Independence. William Swain took this route. Forty-niners striking out from Kentucky, Ohio or Indiana invariably took the Ohio and Mississippi rivers to St Louis.

The principal route beyond the Missouri followed varia-tions on what was known as the Oregon–California Trail. This would take the 49ers through what we today call Kansas, Nebraska, Colorado, Wyoming, Idaho, Utah, Nevada, Oregon and into California. It was a truly epic journey by any standards – and Sarah Royce quickly realized just how tough it was likely to be.

After just four days out, she wrote: 'The sloughs were very bad, stopping us repeatedly during the day, and just at dusk we found ourselves fast, in a most dreary swamp. We had encountered in the middle of the afternoon a tremendous blow and rain while out on the open prairie – the night looked threatening, and before morning we were visited by a heavy thunder storm.

'The next day, Friday, was so inclement as to prevent travelling. I cooked as well as I could by a log fire in a strong north-east blow. My little Mary, to my great surprise, was cheerful and happy, playing in the wagon with various simple things I provided for her, singing and laughing most of the time. Saturday morning, though the weather still continued cloudy, we attempted to proceed, but the rain had softened the ground so much that we found ourselves "stuck" almost every half mile.'

It took Sarah Royce's family one month and four days to cover the four hundred or so miles to Council Bluffs, where they would cross the Missouri River into frontier land. They were almost two months behind the majority of the 49ers. Sarah had no way to judge accurately her family's situation at the time, but those so far adrift were in mortal danger. If they did not make up ground, and quickly, their attempts to cross the Sierra Nevada Mountains into California would have to be made in deadly winter conditions, and this would almost certainly result in death.

The mountain range runs approximately four hundred miles north to south, seventy east to west, and offers as obstacles some five hundred peaks that exceed twelve thousand feet, including the highest mountain in the contiguous USA, Mount Whitney, which is slightly more than half the height of Mount Everest.

Among the first people to attempt to cross the Sierra Nevada into California, two years before James Marshall's gold find, were eighty-seven pioneers from Illinois, collectively known as the Donner-Reed party, after George Donner, his brother, Jacob, and James Reed. The word 'Donner' invokes in Americans thoughts of cannibalism, for that is what the party resorted to after becoming trapped in the Sierra Nevada in the depths of winter.

Only forty-eight members of the Donner party reached California. In subsequent years, some admitted to eating human flesh, including that of children; others denied it ever happened, but the evidence suggests that it did. By the time Sarah Royce and William Swain set off on their journey, following a similar route, they would have been aware of at least some of the stories that attended this infamous migration.

At Council Bluffs Sarah found 'a city of wagons', all waiting their turn to be ferried across the Missouri. The average waiting time was about a week.

'The great majority of the crowd were men, generally working men of ordinary intelligence, farmers and mechanics – accustomed to the comforts and amenities of domestic life, and, most of them evidently intending to carry more or less of these agreeable things with them across the plains,' wrote Sarah. 'Occasionally these men were accompanied by wife and children, and their wagons were easily distinguished by the greater number of conveniences and household articles they carried which here, in this time of prolonged camping, were often, many of them, disposed about the outside of the wagon, in a home-like way.'

The Royces were ferried across the Missouri River on Friday 8 June. It was a dangerously late crossing.

It took William Swain and his friends twenty-three days to reach their chosen outfitting town, Independence, Missouri. Their journey was more than a thousand miles as the crow flies, but much of it was by boat through the Great Lakes and the Illinois and Missouri rivers, so progress was much swifter than Sarah's.

The supply of horses, oxen, mules, food and mining equipment in the outfitting towns was plentiful but expensive. Historians estimate that more than half the Argonauts who took the California Trail were farmers, so many already had much of the equipment they would need. For those who did not, demand drove prices high. For four men travelling together, the cost of a wagon a team of six oxen and enough food for six months equated to about five months' average pay each, depending on the means of transport.

An Argonaut had only to mention that he was bound for California for the price of his purchases to double, but at this stage money was not the highest priority; remaining healthy was – and there were many reports of deaths from cholera. At the time, no one understood that the disease was contracted by the ingestion of bacteria in contaminated food and water. With tens of thousands of people using waterways to head west – many hundreds at a time being confined to overcrowded riverboats, where the hygienic preparation of food and drink was often wanting – its spread was inevitable.

In St Louis, the largest city through which many of the Argonauts would travel, two thousand residents and emigrants died from cholera in June 1849. The riverboats that passed through St Louis served as both its source and transmission. On one boat alone, the *Monroe*, fifty-three people succumbed during a one-month period.

There were times when William, who was prone to stomach problems, must have been terrified. Commonly among Argonauts it was said that a person with cholera might feel fine at breakfast, suffer a little diarrhoea at lunchtime and be dead by supper. In a diary entry dated 3 May, William wrote: 'We find the cholera prevalent here in a virulent form. Today I have felt unwell, have had some dysentery and some disagreeable feelings, and slight griping. This afternoon I took two doses of . . . dysentery medicine, at evening took a dose of peppermint and laudanum. Tonight, I feel better; my dysentery has stopped.'

So, he was not dying today, but that was no thanks to the 'dysentery medicine', which was mostly useless, if not downright fraudulent snake oil.

Perhaps realizing how far out of their depth they were, William and his friends contributed their wagon and team and a hundred dollars each to become members of a company called the Wolverine Rangers, a group of sixty men from Marshall, Michigan. This entitled them to the services of knowledgeable guides, the provision of good food and military-style decision making. Travelling this way was arguably the safest and most efficient.

In a letter to his brother, George, William described the company's members as: 'Americans, mostly eastern and some western men, but mostly smart and intelligent. There are among them two ministers and two doctors, one of whom is said to be well educated and very successful in his practice. There are also blacksmiths, carpenters, tailors, shoemakers and many other mechanics. They are men of good habits and are governed by the regulations of civilised life.'

Within a few days of this letter the number of doctors had been reduced from two to one. On 12 May, William wrote: 'Our company doctor [Palmer] has got the cholera very bad. He is out of his head this evening and will probably die before morning.'

Two days later, his diary entry reads: 'This morning Dr Palmer is dead. Died at three o'clock. There is a solemnity, or rather gloom, on all countenances in the camp. Dr P was buried at ten o'clock in the burying ground on the top of the hill. It was a solemn sight to see one of our number carried to his last resting place far from home and relations. Only thirty-six hours ago he was joyful and mirthful with bright hopes, glowing in his prospects. His relatives little think he is no more. What sad hearts will be at his home when the news reaches them.'

William Swain and the Wolverine Rangers set off from Independence, bound for California, on 16 May 1849.

8

Ken Phelps and the
Deadly Beam of Light

The worst part of me pulled some chicken from a drumstick and took a slug of beer while everyone around the table fell into impressed silence.

'You went out with Dave McCracken?' asked Tom. 'You went looking for gold with Dave Mack . . . *today*?'

'McCracken? McCracken? You went out with Dave Mack?' asked Craig. Then, after a long pause, 'What does he look like?'

I shrugged and pretended to have a mouthful of food.

'Oh, he's some big guy who struts around finding gold,' said Marcy, feeding some meat to her dog, Millie. 'Everyone follows him like he's some kind of god. They all come here with their big RVs and all their equipment, convinced they're going to get rich, and then you see them with their tiny bottles filled with water so the gold in them looks bigger. Crazy fools.' She stopped talking and petting Millie and looked across the table at me disapprovingly. 'Oh, please, don't say you've gotten gold fever too.'

It was a beautiful evening in Happy Camp and I was being cheerfully bitten by clouds of mosquitoes. We were sitting around a long wooden table, surrounded by roses and blossoms on the lovely lawn of the Big Foot RV Park where tents were supposed to be situated. There was none there right now because the only camper actually staying on the site had erected his tent in the section reserved for RVs and caravans.

'Would you like me to move it?' I had asked Rita and Gary King, the owners, earlier in the evening. 'I really don't mind.'

Gary said nothing, giving an impression of unflappability, as if it would take a much more serious problem than this to warrant a reply. Rita had snorted and laughed, before saying, with heaps of sarcasm, 'Oh, don't worry about us and our little rules. You just drop your tent down anywhere you want.'

Gary was quiet and diffident and could often be seen floating between chores and locations with a calmness that seemed to make the air shimmer around him. Rita was loud and funny and used irony as a special power in much the same way as Superman could fly. She was clever and spoke quickly, cracking jokes and telling wondrous stories with total recall and immaculate delivery.

Rita had bought the trailer park for Gary's birthday one year while he was away on business. They ran a TV production company specializing in evangelist shows and this purchase was a validation of Rita's belief that God really did love her. After stumbling into Happy Camp during a vacation some time earlier she had begun buying up houses, abandoned as the region's saw mills fell into decline and closed up. But this was something else, a real leap of faith. The camp – which she had not seen before buying – was notorious as a rundown home for drug addicts, alcoholics and criminals on the run, but she had believed that she could turn it into a haven of calm beauty, and she was right. It is utterly lovely.

When Gary had returned from his business trip, she'd met him at the airport.

'I picked him up, gave him a big kiss and told him I had a surprise for him,' said Rita. 'When we got to Happy Camp, I pulled into the grounds and he just stared at the rundown trailers, rusting cars, all the trash and tenants outside fighting, and he went quiet.

'Then he looked at me and said, "Oh, honey . . . you shouldn't have." '

It was a shame I was in the wrong section; it really was enchanting on the campsite, but I was enjoying Tom's company and so decided I would prefer to remain next to him and his condemned RV, a vehicle that I had named Priscilla. Rita had given her blessing for me to stay there, wandering off with a broom in her hand, cloths hanging from the pockets of her pinafore, yelling, 'Oh, don't mind me. I only own the place . . .'

I was to find that on any given night there would be a weird and wonderful cross-section of characters at Big Foot, and tonight was no exception. Tom, Craig and I had joined Marcy at the long garden table and finished our meal while campers and miners came and went.

I was about to make more of my status as friend and confidant of the legendary Dave Mack when a handsome Native American walked up to Marcy and began chanting. Someone whispered that he was a shaman called Bryon. Marcy appeared embarrassed but raised a smile while Millie tried to hide under her skirt. Marcy was far more relaxed than her pet, the effect of her evening's intake of tequila. I was convinced that she also had a supply of hash cookies, but whenever I raised the subject she would giggle and deny it. One night Craig made a joke about crack cocaine and Marcy, arriving at the punchline, said, 'Ooh, have you got any?'

Shaman Bryon's chanting had the effect of encouraging a crazy-eyed and rambling interloper in our midst to speak louder. No one knew who the man was or from whence he had come; he simply sat down next to Tom, stealing my Dave Mack thunder while telling us about the North Koreans massing on the border with Mexico.

'There's millions of them and they've been there for months.' he said. Craig looked at me and blinked. 'They're getting ready to invade. They've already dropped two atom bombs on San Diego but the bombs didn't go off.' He raised his voice as the shaman chanted louder to be heard above him. 'One of the North Koreans fired a shot into a nuclear power plant and the round went through a cooling pipe and the reactor nearly went critical. The whole of California would have been destroyed if they hadn't fixed it. Doesn't matter though, because the North Koreans'll be here soon, and then we'll all be dead.'

Crazy Eyes stopped shouting long enough for another camper, a large man who I suspected tucked his vest into his underpants, to yell at the top of his voice how he had taken his car to be repaired that day only for the garage roof to collapse on it. We spent the next twenty minutes debating whose insurer would be responsible, while Crazy Eyes insisted it wouldn't matter once the North Koreans came.

In blissful ignorance of this, due to his deafness, was Ken Phelps. Ken was eighty-seven and had brought over his collection of prehistoric shark teeth and the Bronze Star he had earned during the Second World War when he saved the lives of hundreds of men with a beam of light. The shark teeth were impressive in themselves, having been found by Ken a thousand miles inland, but it was his war story that had me on the edge of my seat, trying to ignore the rising cacophony from Crazy Eyes, Shaman Bryon and Roof Man.

Ken, now a slightly stooping but still tall and terribly dignified man, had been a nineteen-year-old electrician third class when the vessel on which he was serving, a troopship called the

USS *Botetourt*, was sent into the Battle of Okinawa during the closing stages of the war.

'We'd just put 2,500 men ashore and the fighting was raging,' he told us. (In all, the battle raged for more than eighty days.) 'The Japs knew they were beat and so they brought in two aircraft carriers with about eighty suicide planes – kamikazes – and the planes were crashing into ships all around the fleet. There were vessels on fire everywhere. I had been told by the chief electrical officer to man a searchlight to look for our downed fliers,' said Ken.

He had a wonderfully endearing way of speaking. If you closed your eyes you might think Jimmy Stewart was telling his story. He was bespectacled and wore a baseball cap emblazoned with the words 'World War II Veteran' above a well-pressed blue checked shirt. He went on, 'These were arc lights about three feet across powered by the kind of carbon rods that welders use. They had a concave mirror behind them and they were so strong you could pick out a pilot in the water five miles away, but I was told I had to ask for permission before switching mine on.

'Suddenly, we were hit by one of the kamikaze planes. It exploded into one of our guns, killing eight men in an instant. I jumped behind the smoke stack and wasn't hurt but the noise was so loud that I lost my hearing; I never got it back. I looked up and saw another plane coming for us and I just reached for the searchlight and flicked the switch. It was instinct – I was wondering if I could blind the pilot.

'Well, he missed the ship and went down into the water. Then I saw another coming, and I pointed the light at him and he went down too. Then another went down. In all, I brought down five aircraft with my searchlight.'

When the kamikaze pilots stopped coming, Ken was summoned to the captain's quarters and made to stand to attention in front of five officers, including the skipper.

'Son,' said the captain, 'who gave you permission to shine that light?'

'Nobody, Sir,' replied Ken. 'But I didn't exactly have time to call the bridge.'

The officers chuckled at the nineteen-year-old's insubordination, and then the captain stood up and approached him.

'He shook my hand and said, "Well done, son. If it wasn't for you we'd be on the bottom now." Then the captain radioed William Halsey, the Admiral of the Fleet, and told him what I'd done, and the Admiral issued orders for all ships to use their searchlights to blind suicide pilots.'

Ken's skipper stated for the ship's record that he would recommend the young electrician for a Bronze Star, America's fourth-highest military honour, granted for conspicuous acts of heroism and bravery, but the USS *Botetourt* was mothballed after the war, its records put into storage – and everyone forgot about Ken. Over the years, he would tell the story of how he brought down five Japanese aircraft with a beam of light, but my guess is that some people might have found it hard to believe. And all the while, for decades, he endured flashbacks and nightmares about kamikaze planes flying towards him in the dead of a massive Pacific night, the screams of dying men piercing the darkness.

'They call it post-traumatic stress these days,' Ken shrugged. 'But back then you would tell a doctor about it and they'd say, "Don't call us, we'll call you."'

About sixty years later, the Veterans of Foreign Wars campaign group was alerted to Ken's case by two of his friends, Bill and Marge. After much toing and froing with a recalcitrant US Department of Defense, the *Botetourt*'s log was found and, sure enough, there was the captain's recommendation. The Department of Defense accepted that Ken was indeed entitled to a Bronze Star and, with breathtaking insensitivity, it invited him to buy one. Finally, the bureaucrats relented and he was awarded the medal gratis.

'One day I got a call from the bank and they said, "Mr Phelps, we'd like you to come down and explain something for us,"' he smiled. 'I went with Bill and Marge and there was fifty thousand dollars in my account. Been put there by the Department of Defense. You get a small annual award with the Bronze Star and it seems that amounted to back pay. From now on I'll get 3,100 a year for life.'

Ken had reduced everyone to silence. I could see that Crazy Eyes was on the verge of saying something about the North Koreans but I gave him a withering look and he shut his mouth, allowing Ken to savour the moment.

The next morning I headed off down the Klamath with my pan, bucket, shovel, trowel, classifier, pick, snuffer bottle, vials, tweezers and hat, all brand new and shiny, for my first attempt at prospecting on my own. The day was stupefyingly hot, something for which I was at first grateful, as it removed the particles of damp that had clung to me when I woke up. Now I wasn't grateful at all; it was 111 degrees.

Given that I had the appearance of the novice skier arriving for his first day on the piste, with pristine equipment but no idea how to utilize snow as a means of transport, I was rather anxious not to bump into any other miners, something that, given our isolation, should not have been too difficult. But just to be sure, I drove way past Independence Camp looking for a likely spot where once upon a time some monstrous flood might have carried down gold deposits from the mountains, which by dint of gravity and the laws of motion would have fallen to the ground and been buried, waiting to be found by me.

How the gold got there in the first place might surprise you – it came from outer space. I found this as hard to believe as you probably do right now, but for three decades the vast majority of geologists have accepted a theory called the 'late veneer hypothesis', the gist of which suggests that some elements, such as gold, platinum, cobalt, nickel and palladium, were dumped here by gigantic meteor showers about four billion years ago. The hypothesis doesn't say that the earth had none of these to start with, but that as the planet took shape, some four and a half billion years ago, they melted and were drawn down towards its molten iron core. Therefore, we shouldn't find them anywhere near the surface, but we do.

According to the late veneer hypothesis, the reason for this is the arrival of millions of meteorites crashing on to the earth after the core had cooled, seeding the outer mantle and the twenty-five-mile thick crust that surrounds it with metals, carbon, water and amino acids, the building blocks of life. The theory doesn't seem so crazy when you consider that in 1998 the spacecraft NEAR (Near Earth Asteroid Rendezvous) identified one asteroid, named Eros, which was thought to contain more gold than has been mined in all earth's history. And, of course, there are billions of them out there.

How the gold got from the crust and upper mantle to where I was now was much easier to accept. Two hundred million years ago, the Sierra Nevada was part of an inland sea. One hundred million years ago, volcanic activity forced the ocean floor upwards in violent convulsions that gave birth to mountains into whose folds and crevices molten granite, gold and quartz were forced by geothermal pressure. These cooled, becoming the rich veins of what California's miners today call the Mother Lode, an underground network of gold-bearing quartz more than 120 miles long that cuts through the foothills of the Sierra Nevada. Then followed millions of years of erosion by earthquake and glacier. Sections of quartz near the surface were exposed by this erosion, while flood and rain washed gold particles and nuggets down mountainsides and into the rivers whose ferocious currents had carried them all the way here as placer deposits.

Now all I had to do was find them.

I felt terribly excited. I had all the windows of my four-by-four open and was struck by the thought that this air, smelling of spruce and pine and oak, might have been the purest I had ever breathed. As I slowed and swerved, looking for my perfect place, I realized that much of the river was inaccessible without ropes or a boat. (Why had I not brought ropes? Surely no adventure was complete without ropes?)

I slowed for a raccoon and her cub to pass before veering towards a crow pecking at some roadkill. After forty minutes driving I had failed to find the right spot. In truth, I did not know what I was looking for, so I turned and headed back to an easily accessible area named Wingate Bar that I had sighted closer to Happy Camp but which I had rejected for that very reason. I hadn't seen a soul all morning.

I unpacked everything self-consciously and loaded it into my rucksack and bucket before adding several bottles of water, some mosquito repellent and a sandwich and heading down a faint path towards the river. This was it, I thought, I'm prospecting for gold! I had rarely felt sillier and found myself looking over my shoulder for bears or miners who might see me waddling, green as cabbage.

I padded through dogwood and river grasses, catching the smell of far-off honeysuckle and looked up to see a bald eagle and a passing dove, one riding eddies effortlessly, the other rushing as if on important business. I encountered my first poison oak and gave it a wide berth, its presence reminding me suddenly to make enough noise to alert snakes and bears to my presence so they might slope away without confrontation. I rattled my bucket.

After a few minutes, the trail opened into a clearing above the river and the silence was shattered by the sound of a loud engine. Below me was a miner in a very large hole. He had clearly been working this spot for some time; he was about thirty feet away from the river and he had dug into the bank, about ten feet above the waterline. I was later to find out that this was called high-banking, for obvious reasons. It involved digging into soil that was not habitually covered by water, but which would have been inundated during the kind of flooding I was looking for. The miner would dig in front of rocks or in crevices where gold might have fallen, again trying to get as close to bedrock as possible. Once graded through a classifier to remove unwanted rocks and stones, the miner would run his soil through a sluice box using water pumped up from the river, before panning the pay dirt that was left behind. However, the responsible miner would also dig a pond into which the muddy run-off from this operation would be released so that it would settle before returning to the river as clear, clean water. And that was what this man was doing.

He had his back to me and I was wary of sneaking up on him in case he was armed. I cleared my throat and coughed, then repeated the process more loudly. He couldn't hear me. He had his head lowered into the hole he had dug and was using a vacuum pump and hose to suck up all the loose dirt in the crevices he was uncovering.

'HELLO!!' I yelled.

The miner shot up and turned around, switching off his pump. My ears were ringing.

I advanced with my hand out and he shook it, weighing me up and smiling. His name was Doug McDowall and he was a retired electrical utility worker from Rainier in Oregon. 'That's appropriate,' I said when he told me his name.

'What?'

'Doug.'

'What?'

'Nothing.'

Doug was sixty-two and was wearing a pair of wellington boots, a checked shirt and muddy blue jeans supported by red and black braces. His white-bearded face was shaded by a camouflage baseball cap and he appeared vaguely amused.

'Can I walk here?' I asked, not wanting to give the impression that I was a claim jumper.

'Sure, why wouldn't you?'

'Is this your claim?'

'Well, it's a New 49er claim but while I'm working it, yes, it's mine.'

The club had a rule that while a member was actively searching for gold on one of its claims, the immediate area around their activities was theirs to work exclusively so long as there was evidence of labour. I would subsequently find myself in the middle of nowhere, on the banks of a creek or river and far from any other humans and I would see a shovel or a few buckets that told me someone was working that area. In spite of the enormity of the landscape and its emptiness, I would have to move away from those diggings and find another place to work. In turn, the person with the claim had a duty to fill in any unsightly holes and return the site as close as possible to the condition in which they had found it.

'You finding much?' I asked, and as soon as I did I wondered whether this might be considered bad form, too personal a question, but Doug didn't seem to mind.

'Been doing alright but I think it's pretty much played out now.' He surveyed his handiwork and scratched the back of his head. 'Gonna have a big job putting this right.'

I could see the miner looking at all my new equipment.

'It's my first time,' I said. Doug raised his eyebrows. 'This moment is literally the first time I have ever tried to find gold. Right here, right now.' His eyes flitted to my bucket. 'Never been used. Look, my pick . . . It's shiny.'

'Sure is,' he said. 'You know what you're looking for?

'Well, I've read stuff and I've seen some guys with a dredge.'

'That won't do you any good here. C'mon, I'll show you.'

And this wonderful man put down his tools and demonstrated why he was searching in this place (because there had been a flood in 1996 and there was evidence of its terrifying presence here, at a spot where rocks and a bend in the river would have slowed it down). He told me to search crevices and folds in bedrock and to empty them with purpose because I could be sure that weather and water would have worked with gravity to drag the gold deposits down there.

'You know how to pan for gold?' he asked.

'I've seen it done. I get the principles.'

'Well, if you want any tips, you come and ask.'

I looked downriver and asked Doug if he would mind if I tried my hand about two hundred yards away. He said he wouldn't mind at all; his claim did not extend that far. I thanked him, scooped up my equipment and trudged down the bank, climbing over rocks, pushing poison oak aside with a stick and whistling loudly to discourage bears. Once I was out of sight of Doug's claim, I dropped everything and climbed the bank, looking for evidence of the old flood – and I thought I found it; rocks that seemed out of place high above the water, deep granite creases that were filled with sand and pebbles that had travelled far from the normal course of the river. I decided to clear out the rocks and dirt from a crevice about two feet wide, where I imagined water crashing, slowing down and losing the gold that had been in thrall to it.

It didn't occur to me in any significant way at the time – at least not consciously – how stupid it was of me to ignore the fact that I was on the outside of a bend in the river and that on its inside was a gravel bar that would have been a much more intelligent place to prospect. I vaguely remember registering this fact and immediately dismissing it on the basis that if I were to act on it, I would have to wade across the river and get wet.

I began to dig and realized that I had never put a shovel to any productive use before. I had grown up with a garden, but it was the preserve of my parents. As a child, I never built sandcastles, preferring instead to go exploring in the treelines that almost always lay behind the dunes. And as an adult, living mostly in London, where space came at a premium, I had never had a garden, always repairing instead to one that came with beer. Finding now that the most basic requirement of gold prospecting – the digging – was new to me was desperately embarrassing, as if embarking on a marathon with the Damascene realization that you didn't know how to run.

I set aside several big rocks that covered the crevice and then dug deeper and deeper, to a depth of around three feet, where I could feel some kind of bedrock, before throwing the soil and sand on top of the classifier sieve wedged in the bucket. I then shook the classifier until the smaller particles fell into the bucket. I examined the remaining rocks and pebbles in case they contained a nugget, but they never did. I cast them aside, repeated the process and when the bucket was half full I rushed down to the river with it to begin panning for the gold that it surely contained.

I could not imagine a more wonderful way to spend time or a more beautiful location in which to while it away. I waded into the water and sat down on a rock, my toes digging into the cool pebbles and soft green weeds, and marvelled at the loveliness in which I was carrying out my labours. The river was only thirty feet or so across at this point and on the other side was a flat clearing with thin growths of willow and young white fir. To the south, the Marble Mountains, draped in spruce and ponderosa, reared up to the sky. Where the bend arrested the river there was a deep pool. Sweat poured down my face and I resolved to swim in it later.

I put a trowel-full of pay dirt from my bucket into my pan and submerged the whole thing into the river between my ankles, shaking it vigorously so any gold would find its way to the bottom. Then, remembering how I had seen Rich and Dave Mack do this, I carefully washed away the surface layers, shaking and twisting the pan regularly so that the heavier deposits would be teased downwards. My excitement was reaching a peak: I would soon see gold, just as the extreme prospectors had. Eventually, all that was left was black sand, heavy with iron, and I told myself that the gold should emerge any second as I gently washed it away.

But there was nothing.

I washed everything in the bucket but found no gold. It was the same with the second bucket. Then, in the last pan from the third bucket, my hole in the ground having grown waist-deep, sweat coursing down my brow, my hands and knees raw, I was suddenly dazzled by the appearance of a small flake as it emerged from the black sand in my pan. There was just one sliver, about the size of half a grain of short-grain rice, flattened to the thickness of a human hair, but it was unmistakeable and in that moment I imagined it lit up my face.

It was gold, and it was all mine.

9

The History of Yellow

The first gold flake I found at Wingate Bar was the only one I uncovered during a long and exhausting day, but that didn't worry me too much. I was learning and it would be only a matter of time before I would find riches beyond my wildest dreams.

After I had unearthed it, I floated on my back in the river, following the eagles, breathing in the mountain air and thinking about gold and humankind's love affair with it. It made little sense to me.

As a metal, gold is fairly useless. It is too soft with which to build anything or to forge into effective weaponry; a good steel blade would simply slice through it. Because it is inert, it doesn't react with other elements or chemicals and so you can't make gold do anything interesting. It just sits there for all eternity, unchanging, as if saying, 'Look at me, aren't I beautiful?'

It was these very qualities, however, that made it desirable to early humans; it was shiny and had a pretty colour – its very name comes from *geolu*, the Old English word for 'yellow' – and it could be beaten, without exposure to heat, into attractive shapes as jewellery or decoration. Because it was scarce, if you had some then you were someone to be respected. If you had lots of it, you were someone to be feared. Its agelessness aligned it with vitality and longevity, and that, too, made people desire it. The Roman philosopher Pliny the Elder described its use as a cure for ailments ranging from ringworm to piles, though if it acted in any way on these conditions, it must have been as a placebo.

The earliest people to mine gold in any quantity were the Egyptians, some six thousand years ago. They called it 'the flesh of the gods', and at first only kings were allowed to wear it. They mined it in the Eastern Desert and Nubia, straddling the border with northern Sudan, and they extracted it ruthlessly, enslaving thousands of people at a time to mine it for them, people whose sole purpose was to dig for gold ore and then die.

Describing conditions in the mines of the Egyptians, the Greek historian Diodorus Siculus wrote in 60BC that prisoners of war, criminals and innocent members of their families were all condemned to extract gold. They were beaten and starved, and often poisoned as a result of the practice of exposing gold-bearing quartz to fire in order to weaken it, a process that gave off deadly arsenic fumes. Legions of them died from their labours. Siculus wrote:

At this task they labour without ceasing beneath the sternness and blows of an overseer. And since no opportunity is afforded any of them to care for his body and they have no garment to cover their shame, no man can look upon these unfortunate wretches without feeling pity for them because of the exceeding hardships they suffer. For no leniency or respite of any kind is given to any man who is sick, or maimed, or aged, or in the case of a woman for her weakness, but all without exception are compelled by blows to persevere in their labours, until through ill-treatment they die in the midst of their tortures.

In 2001, a group of German geologists and Egyptologists published a report in the *Journal of African Earth Sciences*, which came to the conclusion that all of this suffering under the pharaohs resulted in the production of less than eight tons of gold. Their findings ran counter to all the other assumptions made by historians about Egypt and the country's mountains of precious metal. Previously, the pharaohs were thought to have extracted much more of it. The British numismatist C.H.V. Sutherland, in his 1959 book *Gold: Its Beauty, Power and Allure*, estimated Egypt's output at 670 tons between 4000 and 2000BC alone.

In their unexpected report, the geologists wrote:

The wealth of gold in Pharaonic Egypt is legendary, as illustrated when in about 1340BC a Mesopotamian Mitanni ruler asked Pharaoh Amenhotep III, in an urgent letter, for a larger gold consignment, arguing that 'gold occurs in Egypt like sand on the roads'. After our inspections of nearly all of the many known Pharaonic gold production sites in Egypt and NE Sudan, all these imaginative expectations must be rejected.

During the entire Egyptian history of roughly six thousand years' tradition in gold production, less than the monthly gold output of present South Africa was achieved.

In 2001 that would have been about thirty-three tons. In an almost apologetic tone, the academics concluded: 'This estima-tion, of course, does not correspond with the expectations of archaeologists when projecting the gold inventory of untouched tombs, like the one of Tutankhamun, and inferred from royal dedication lists in temples from the entire Pharaonic history. The only solution to this discrepancy might be seen in intensive recycling of gold more or less constantly throughout the Pharaonic history.'

With the decline of the pharaohs came neglect of the ancient mines, but after almost two millennia they are producing gold again from the eastern deserts. The Sukari mine, the country's largest, is expected to yield half a million ounces a year from 2015. Mining experts believe that Egyptian reserves could amount to some seventy million ounces; the pharaohs had been sitting on it the whole time.

Once humanity took the first steps from localized subsistence to more widespread production and trade, gold began to come into its own. Though dense, it was portable and could be bartered for other goods because it was in demand. Everyone wanted some, and that made it universally useful as a means of exchange.

As early as 4000BC the pharaoh Menes was casting gold into bars, but more as an index of power and store of wealth than as money. For gold to be useful as currency, those employing it needed to have confidence in its purity and weight, and this required a system of measurements accompanied by a whole new lexicon from 'carat', denoting purity (with twenty-four carat gold being completely pure, fourteen carat being fourteen parts gold and ten parts another metal, and so on), to 'grains' denoting weight. The carat was originally used as a measure of weight rather than purity, with a carat being the seed of the carob tree (*Ceratonia siliqua*), each one of which, by a remarkable quirk of nature, weighs exactly one-fifth of a gram.

The grain, which was reputedly first employed as a weight by the merchants of Troyes in north-eastern France, refers to a grain of barley, which by a similarly natural quirk of nature always weighs the same, no matter the size of the ear from which it came; there are 480 of them in a Troy ounce of gold. Whenever prospectors refer to an ounce of gold, they mean a Troy ounce, which is three grams heavier than a regular *avoirdupois* (from the Anglo-Norman French meaning 'goods of weight') ounce, and, therefore, harder to amass.

As a means of exchange, gold remained unreliable until the discovery in the Iron Age kingdom of Lydia, in modern-day Turkey, of a dark stone that came to be known as a touchstone. When gold was rubbed against this stone, it left behind a streak that varied in colour depending on the purity of the gold. The touchstone allowed for more accurate exchanges until the first minting of coinage, which, according to the Greek historian Herodotus, took place during the reign of King Alyattes, also in Lydia, sometime between 610 and 550BC.

Alyattes's coins were made from electrum, an alloy of gold and silver, but they were replaced with coins of pure gold and pure silver by his son, Croesus – who retained the basic denomination that his father had introduced, the stater, and subdivided it into twelfths. The Lydian coinage was so well-made – and so reliable was the content of its precious metals – that it quickly became trusted throughout Asia Minor, Greece and the rest of southern Europe as the currency of choice.

It also meant that gold was now more practical than symbolic as an indicator of wealth.

After the success of Lydian coinage, it became de rigueur for powerful kingdoms and empires to mint their own and to update their designs regularly, not least because when this happened, the king, queen or emperor took a tiny cut of each – if the ruler's head was on the coin, then it could be trusted and was worth more than the value of the actual gold it contained. The difference went to the monarch as *seigneurage*, denoting his or her sole right to mint coins. The penalty for attempting to forge or trim valuable metal from coins was usually death.

The reliability and purity of a nation's coinage – usually gold, silver and bronze – became a matter not just of pride but also power. Where a coin was trusted above others, its use became international, spreading far and wide the name and reputation of the king who minted it. The Persian Emperor Darius, Alexander the Great, Julius Caesar and Constantine the Great of Byzantium understood this and minted accordingly.

Of course, in order to keep minting your coins, you needed constant supplies of gold and silver, and the best way to secure these supplies was to conquer territories known to have productive mines and to take them for yourself (gold has been found on all seven continents, so there was plenty of opportunity). This economic model fuelled war on an almost continuous basis, not least because the by-product was a constant supply of prisoners who could be used as slaves to unearth precious metals for the victor (and then die). And the more countries you conquered and gold you grabbed, the more coins you could introduce to a larger area, stimulating trade and making you wealthier and more able to conquer the next country.

Probably the most successful gold coin in history was the solidus. This was first minted in Byzantium by Constantine the Great in the fourth century. Variations on it, collectively called the bezant, facilitated international trade for around a thousand years.

In comparison, the dollar and the euro might be regarded as fledgling currencies.

So, I had found gold and my spirits soared, but one thing weighed on me and it was to do with weight itself. I had been using a five-gallon bucket, which could hold water weighing about fifty-five pounds. I had read that dirt was about three times heavier than water. That meant each bucket of soil weighed about 150 pounds, and I had classified and panned six half-buckets (I could barely carry a whole one).

I had dug and processed about 450 pounds of dirt – one-fifth of a ton – and found one speck of gold. What kind of a task had I set myself?

At the RV park that evening, Marcy caught me showing Tom the gold flake in my vial and called me a damned fool. 'You'll be putting water in there next to make it look bigger,' she said.

One of my more successful purchases at the start of the trip had been a battery-powered lantern that cast a soft red glow and it was in the warmth of this that we sat, Tom, Craig, Marcy and me. Craig was not there to prospect – I felt he was just hiding from something – and he found my growing obsession amusing. Tom was passionate about gold, but his determination to find serious amounts of it meant that he had found none at all.

He had arrived from Australia two months earlier, having read as much as he could about dredging, which was outlawed in his adopted country. He figured if he could no longer dive under the sea, he would dive under white water and look for gold. Using the internet, he had lined up second-hand equipment in California and resolved to buy an RV once on the ground. But it all went wrong; he ended up travelling hundreds of miles, as far south as Sacramento and as far north as Portland, looking for the perfect vehicle but he didn't find it. He hired a car and spent hundreds of dollars on gas and motels along the way. Finally, in Klamath Falls, he was offered Priscilla, a relic from the 1970s, and it broke down almost immediately. He replaced all the brakes, then the tyres, then some of the electrics and indicators. By the time he arrived at Happy Camp, he had spent most of the money he had set aside for his trip.

To make matters worse, once he saw the New 49er claims, Tom realized that without a boat or a prospecting partner he could not get his dredge down to long stretches of them. It was too heavy to haul there by himself. He raised this problem with Dave Mack, but found the extreme prospector unsympathetic. This had upset Tom, who was a sensitive man. But it had not upset him nearly so much as the email he received from his wife during my first week in the camp, which accused him of being a failure.

The message was probably sent in haste, in a lonely fit of pique perhaps, and I did not hear anyone judge his wife harshly for it, but it crushed Tom and we all noticed it immediately. He was bombarded with kind words for no reason other than that he deserved them. I noticed from the start how Tom was held in high esteem by the other miners, not least because he was often to be seen unobtrusively helping them.

On my second afternoon in Happy Camp, I found a duvet in my tent; inquiries revealed that Tom had put it there after overhearing me describe to Terry the plummeting night-time temperatures. He was always helping Marcy, shopping for her or just escorting her the twenty feet to her RV at the end of each evening. He would notice the first signs of her preparing to retire and his attentiveness towards her was appreciated by Marcy and admired by us.

Whenever I asked Tom about his life he seemed always to be putting a positive spin on unlucky events. A broken relationship in Scotland many years earlier had driven him to Australia. Two children were left behind and this seemed a source of sorrow or regret. Every time he seemed to be doing well, something went wrong for him. For a time he had made a living diving for abalone in Tasmania and was enjoying a measure of financial success before coastal police caught him without the proper licence to catch and sell the mollusc. It turned out that this was such a big deal in Tasmania that he was ordered to pay a fine of sixty-seven thousand Australian dollars (sixty thousand US dollars). Later, when he missed some payments, he was thrown into prison for twelve days.

Tom was in almost constant pain, physically, from his back injury – he had three herniated discs from a lifting accident – and, emotionally and financially, from having to give up his diving career. He maintained good posture in spite of his suffering, and was handsome for his sixty-five years, with white hair and a suntanned, lined face. He had sallow cheeks that told a story, again to do with bad luck. Some time ago he had paid a considerable sum for extensive work to give him beautiful teeth, but of all the dentists he could have chosen, Tom's was appallingly negligent. All the work collapsed and fell out, leaving him with only stumps that he could not afford to put right . . . unless he found some gold. But he had found none here and a brief sojourn to the Rogue River in Oregon had proved fruitless too. Nothing, it seemed, would go right for him.

The day after he received the email from his wife, Tom did not emerge from his RV. He'd had to move to a different part of the camp because his plot was needed for a modern vehicle with electrical and plumbing connections that were decades ahead of Priscilla's. It was Tom's luck to be placed next to a family whose child screamed all night – a child who went uncorrected and whose father thought it appropriate to saw wood and hammer and bang random things against one another from 5am each day.

I fetched some coffee from Parry's Market and knocked on Tom's door. He opened it.

'You okay?' I asked. He looked terrible. I had seen dead bodies with healthier complexions. He was shivering in the morning heat.

'I'm off the oxycodone,' he said, taking the coffee. 'I've been wondering if it's affecting my judgement, if it's why things keep going wrong. And the email . . . well, that made my mind up. I've got to go cold turkey.'

I didn't know what to say. Coming off legally prescribed oxycodone, which is opiate-based, can be as tough as breaking free from heroin. I just nodded and asked if he needed anything. He crawled back into bed and I said I would call in on him later.

I headed off to Independence Camp and drove over the bridge that seemed to go nowhere in order to try my hand at panning again. (In fact, the bridge joined an old logging trail that had fallen into disuse with the slow demise of the lumber industry.) I recalled Dave Mack saying that some miners had been finding decent deposits on this side of the river, so I walked downstream looking for a likely place to dig. I saw no one else working, though some high-banking equipment had been left at one spot, which I dutifully passed.

After about half an hour I found a grove shaded by willows and peppered with four-foot-high rocks at the foot of a steep hill that might have captured gold during past floods. No one had disturbed the ground, so I set up and prepared to hack into the baked mud. The temperature gauge in my car had told me it was ninety-eight degrees.

'Hey,' said a voice from behind.

I almost leapt out of my skin, reasoning with lightning speed that bears couldn't talk but feeling as unsettled as if one had. I spun round to see a big man in a straw hat and a blue Hawaiian shirt that failed to cover his downy and mosquito-bitten belly. He had rusty candyfloss hair and an unkempt beard that fell appealingly round a gummy mouth that seemed unwilling to stop moving.

'Hey yourself,' I replied. I took a few breaths and extended a clammy hand.

'Bill,' he said. 'Most folks call me Wild Bill.'

Wild Bill!

'Why do they call you that?' I asked, smiling. He didn't seem too ferocious.

'Don't know,' he shrugged. 'Guess they think I'm wild.'

Wild Bill, who I had put at about forty years old, was carrying his equipment in a small backpack. He had a staff that he leaned on when he walked or stood talking.

'You got a stick?' he asked.

'I've got this little thing.' It was a small branch from a fallen poplar.

'Well, that's no good. You need a big stick, and you keep hitting the floor like this, let the rattlers know you're coming. You digging here? No good here. You need to go farther downstream. I got a good place down there and it keeps paying out, but you gotta go some distance, put some space between you and anyone else who's on the river.

'That your pan? This is better.'

My pan had small riffles that went against the outward flow of water to catch the heaviest deposits – gold. He pulled out a smooth tin pan that had no riffles and which I knew was for more experienced prospectors than me.

'And you should get one of these.' He held up a small aluminium funnel that he said could classify much quicker than my equipment. I needed to pass my dirt through mesh first and into a bucket before grading it further; his funnel did the job all in one go.

'Thanks,' I said.

'You going down to bedrock?'

'Yes.' This, at least, I knew I was doing right.

'Well, don't. Best deposits here are six inches above bedrock.'

'Really? Thanks.' There was a brief silence while Wild Bill looked at the river.

'You hot? You look hot,' he said.

'Yes,' I nodded. 'I'm not used to this heat.'

'Well, this is the most important lesson I can give you.' He put down his staff and took a step towards me. 'Give me your hat.'

I handed my straw hat to Wild Bill, wondering whether he might not be the most helpful stranger I had ever met. I smiled quizzically at him. He seemed very concerned about every aspect of my equipment, welfare, location and experience. Then he walked down to the river and dunked my hat in it.

'Here,' he said, holding out the sodden headgear. 'Air-conditioning.'

I put it on and experienced the most wonderful sensation, not only from the cool water running down my face and neck – that was a given – but also from the breeze rising from the Klamath that passed through the brim of my hat and turned to chilled wispy fronds of coldness that clung to my scalp. I instantly felt ten degrees cooler.

I realized I was laughing. 'Thank you,' I said.

'Well, good luck,' he nodded, whistling as he tapped his staff and disappeared into the undergrowth.

Two hours later, having uncovered a small but thrilling number of gold flakes, I hiked back to my car and found a note scribbled on a piece of paper under the windscreen wiper.

'Come to Independence Camp,' it read. 'We need to talk.'

And it was signed Wild Bill.

10

The Wild, Wild West, Naming of a Town and the Karuk

By the time Sarah Royce and William Swain had crossed the Missouri River and begun the second leg of their journey to California, more than thirty thousand Argonauts were already ahead of them. 'Such an immense number had already left the Missouri, and were far on their way, that the grass was all eaten up, and no more animals could live on the great plains,' wrote Sarah.

This has to be read again to be fully appreciated. The risk of having insufficient grass for your oxen at the start of a four-month journey across virtually unknown territory would be the modern equivalent of attempting to drive across half of the United States in the knowledge that almost none of the gas stations that lay ahead had any fuel, and would not have until the next year. Yet Sarah, William and thousands more like them went on regardless.

Only six weeks after President Polk's gold speech, the *New York Herald* described the unprecedented movement of people thus:

> The spirit of emigration which is carrying off thousands to California so far from dying away increases and expands every day. All classes of our citizens seem to be under the influence of this extraordinary mania . . .
>
> What will this general and overwhelming spirit of emigration lead to? Will it be the beginning of a new empire in the West, a revolution in the commercial highways of the world, a depopulation of the old States for the new republic on the shores of the Pacific? Look at the advertising columns of the Herald or any other journal and you will find abundant evidence of the singular prevalence of this strange movement and agitation in favour of gold digging on the Sacramento. Every day men of property and means are advertising their possessions for sale in order to furnish them with means to reach that golden land . . .
>
> Poets, philosophers, lawyers, brokers, bankers, merchants, farmers, clergymen – all are feeling the impulse and are preparing to go and dig for gold and swell the number of adventurers to the new El Dorado.

When the Argonauts left the outfitting towns they faced two thousand miles of plains, prairies, deserts and mountains. They negotiated sodden grassland, swollen river and stubborn rock, sometimes travelling up to twenty miles a day, sometimes little more than a few feet over scree and mud and bone-jarring crag. Once Sarah and Josiah crossed the Missouri in June – a very late crossing indeed – they casually hooked up with other travellers while William and his friends, three weeks ahead of Sarah, progressed with the Wolverines.

The journey had several distinct zones and landmarks with which overlanders felt strangely familiar, often before even setting out. Several guidebooks to the route were in wide circulation, as were accounts of hardship and success that filtered back down the line like osmosis. Many, like Sarah and William, relied heavily on the accounts of John C. Fremont, 'The Pathfinder'.

First came seven hundred miles over the Great Plains to the Rocky Mountains. The Argonauts travelled through present-day Nebraska upstream along the Platte and North Platte rivers, gently ascending all the way to the Continental Divide that ran down the middle of the nation. This part of the journey was arguably the easiest; long stretches still had plenty of grass for the teams, fresh water and game to hunt as the landscape became bereft of trees. Many gold-seekers saw their first buffalo here and, to the dismay of the Plains Indians they passed, sometimes killed them for fun, leaving carcasses to rot in their wake. There were tens of millions of buffalo before the migration of white people westwards. By 1890 there were fewer than eight hundred.

Native Americans were regarded almost universally as savages. Even the Argonauts I took to my heart – those who thought of themselves as decent and God-fearing – found ways to shock with their attitudes towards the indigenous population. Sarah described her first encounter with them in wild country after seeing dots on the landscape far away. At first she thought they were buffalo, then, drawing closer, she realized they were men.

'They proved to be Indians by the hundreds; and soon they had ranged themselves along on each side of the way,' she wrote. 'A group of them came forward, and at the Captain's command our company halted, while he with several others went to meet the Indians and hold parley. It turned out that they had gathered to demand the payment of a certain sum per head for every emigrant passing through this part of the country, which they claimed as their own.'

Of course, in the minds of the white Americans the concept of Manifest Destiny outweighed any claim to territory that the Native Americans might have. Payment of any kind would be unacceptable – as far as the migrants were concerned, this was their land.

'The men of our company after consultation resolved that the demand was unreasonable! That the country we were travelling over belonged to the United States, and that these red men had no right to stop us,' went on Sarah, forcefully, demonstrating the prevailing mindset. 'The Indians were then plainly informed that the company meant to proceed at once without paying a dollar. That if unmolested, they would not harm any thing, but if the Indians attempted to stop them, they would open fire with all their rifles and revolvers. At the Captain's word of command, all the men of the company then armed themselves with every weapon to be found in their wagons.

'Revolvers, knives, hatchets glittered in their belts; rifles and guns bristled on their shoulders . . . We were at once moving between long but not very compact rows of half-naked redskins; many of them well-armed; others carrying but indifferent weapons; while all wore in their faces the expression of sullen disappointment, mingled with a half-defiant scowl, that suggested the thought of future night attacks, when darkness and thickets should give them greater advantage.'

Many Gold Rush diarists spoke of Native Americans as thieves and beggars; others described Argonauts paying for sex with Indian women and calling it 'otter hunting'.

In common with all the overlanders, William Swain would not have hesitated to use his gun against a Native American attacker, but he at least found it in his heart to admire their culture. In one diary entry at the start of the Plains leg of his journey he wrote: 'It really looks singular here in the woods to see the banks of the river lined with wagons, men and cattle, and Indians riding on fine ponies decorated in the finest style saddles, bridles, martingales and breast straps. Many of the Indians have strings of small bells around their horses' breasts. They wear leggings, blankets or shawls, turbans around their heads and make a grand appearance as they prance around the camp. They are splendid riders.'

Travelling the three hundred miles between the army outposts of Fort Kearny, Nebraska and Fort Laramie, Wyoming, awe-inspiring rock formations greeted the emigrants – Courthouse Rock, Jail Rock, Chimney Rock, Castle Rock and Scotts Bluff. These were all landmarks featuring in the chronicles of earlier explorers such as Fremont, which told them how far they were into their journey. And, of course, how much of it was still to come.

Here in the valley of the North Platte River the going was easy compared with what was to come, and there were times when the overlanders behaved like tourists. If they'd had cameras, their progress would have been halted intermittently as they took photographs.

On 29 June William wrote: 'This morn we have Courthouse Rock behind us, and on the left have many other curious-shaped mounds. But the object of attraction with all of us is the celebrated Chimney Rock . . . This column will not probably stand long, as it is badly scaled, cracked and falling fast. Fremont makes it three hundred feet high. From a distance it appears like the high tower of a steam factory.'

Chimney Rock, in Morrill County, western Nebraska, some four thousand feet above sea level, was indeed as high as John Fremont estimated. Its use as a navigational landmark along the California Trail cannot be overstated. In spite of William's concern for it, it remains standing today.

West of Fort Laramie, the Argonauts entered the Rocky Mountains in the south-eastern corner of modern-day

Wyoming, following the North Platte River along the north-ernmost edges of the Laramie mountain range. Most would have been about a month out from the trailheads on the Missouri River by now. The Wolverines had been making slow progress, possibly because of the size of their party and the number of times they would stop due to illness; Sarah and Josiah crossed into frontier territory almost a month after them but were now barely a week adrift.

The next landmark of significance for the travellers was Independence Rock, situated in modern-day Natrona County, Wyoming, a 130-foot-high lump of granite that sticks out from the flat scrub like a boil on a supermodel's face. It was given its name by a fur trapper and explorer named William Sublette, who celebrated Independence Day there in 1830. It was gener-ally believed by the Argonauts that they would need to make Independence Rock by 4 July if they were to miss the worst of the autumn weather once they reached the Sierra Nevada. William passed it on 20 July, Sarah on 26 July.

Beyond this point, the landscape changed suddenly and for the worse. The routes, such as they were, became strewn with abandoned wagons, furniture, tools and anything that the struggling emigrants believed they could live without in California. The 'conveniences' that Sarah had described around the wagons of those who brought their wives were the first to go. The ground, she observed, was 'utterly destitute of vegetation'.

William wrote of it: 'The features of the country are entirely changed. East of Fort Laramie the bare prairie, naked and level was the feature, but here it is that of a broken, rocky, mountainous country. The broken ledges of bare rock and sparse, scrubby pine rising hill upon hill is the ever-changing scene that meets the eye. Our road has been strewn with articles left by the emigrants to lighten their loads.'

In his intriguing book *Hard Road West* (2007), the geologist Keith Heyer Meldahl explains – most wonderfully – just why the California Trail proved so difficult and why, without a singular incidence of tectonic flukery, there would not be a trail at all. This geological fluke announces itself along the eastern length of the Rocky Mountains as a break in a series of colossal stone uplifts known as the Foreland Ranges.

Meldahl explains:

> The Foreland Ranges face the Great Plains like a great wall. But a wide gap in the wall exists in Wyoming, between the Laramie, Bighorn, and Wind River ranges. This is why the Oregon–California Trail passed through here. By following the valleys of the North Platte River and then the Sweetwater River west through this gap, the emigrants could ascend gradual slopes all the way to the Continental Divide at South Pass.
>
> Whether by God's hand or Nature's, South Pass exists by geologic consent. After the Foreland Ranges squeezed upward, a massive mountain blocked the way to South Pass. Later, as the crust stretched, this mountain slid down like a wedge between two large faults to form the Sweetwater Valley, opening the way west through South Pass. Had this not occurred, there would be an unbroken mountain barricade from Montana to New Mexico, and the Oregon–California Trail – and America's westward expansion – would not exist as we know it.

South Pass, at 7,500 feet, marked the half-way point to California. Here, for the first time, the Argonauts would see rivers flowing west to empty into the Pacific, instead of east into the Atlantic.

Shortly after South Pass, the Wolverines headed due west along a route known as Sublette's Cut-off, named after the explorer, and into punishing mountain ranges and rivers flowing north–south, forming hurdle after desperate hurdle. Their path was strewn with dead animals and even more possessions left by Argonauts facing death in a triage of decreasing necessity. Sarah's family took the more southerly Mormon Trail, where they almost immediately found their path arid and their water alkaline. They made their way to Salt Lake City to recover and resupply, one of their oxen having died, arriving on 19 August.

More than half of their journey was completed, but fewer than half their trials endured.

I was finding the history of Happy Camp to be fascinating and tragic in equal measure. The population of the town – around twelve hundred – was composed of sixty-five per cent white and thirty per cent Native American, with a smattering of other races that had settled there. The Native Americans were from the Karuk tribe, with Karuk meaning 'upriver people'. Their nearest neighbours were the Yurok, the 'downriver people', to the south-west, and the Hupa, 'people of the place where the trails return,' to the south. The cultural anthropologist Alfred Kroeber, in his *Handbook of the Indians of California* (1925), collectively described the three tribes as the Klamath River Indians.

According to the US Department of Agriculture and the US Forestry Service, the town was given its English name by fourteen drunken gold prospectors in July 1851. These miners had travelled north across hard country on hearing of deposits in and near the Klamath. They did, indeed, begin finding gold in Indian Creek close to where it emptied into the river, and they decided to make the location their home. It would be fair to say that the group was in rather good spirits.

The official literature quotes the historian A.J. Bledsoe writing in *The History of Del Norte County* (in which Happy Camp was once administratively situated) in 1881 thus:

> The place owes its name to the fact of a little celebration by the boys in honour of the place in which they had established their future homes. It appears that on a certain evening the whole party were assembled together and having a high old time. The bottle from which they refreshed the inner-man was passing freely from hand to hand. And, through the exhilarating effects of its contents the company was beginning to feel in accord with the spirit of the occasion and disposed to cast dull care away and enjoy themselves while yet they might.
>
> Presently, someone proposed that the place should have a name when one of them, who perhaps was particularly satisfied with their location, suggested that it be named 'Happy Camp' and immediately three hearty cheers were given for the Happy Camp and the bottle was passed again. Thus the new mining town received its christening.

'Ha!' said Rita when I mentioned this account. 'That's what they want you to think.'

'It is?'

'Absolutely. They don't want you to know how it really got its name.'

'They don't?'

'Of course not. The truth isn't so pretty.'

'It isn't? What is the truth?'

'Well, don't tell anyone I told you this, but it got its name because of the Chinese.'

'The Chinese?'

'Yeah. There used to be a lot of Chinese miners but they weren't welcome and they weren't allowed into the saloons so they made their own entertainment with opium. They used to sell it to the other miners and would go up and say, "You want? You buy . . . make you very happy," and so it came to be known as the Happy Camp, because it was home to a bunch of junkies.'

Rita was correct about the Chinese presence; the 1880 census put the population of Happy Camp at 597, made up of 97 Native Americans, and 250 each of white people and Chinese.

Later, I told Terry about this and he laughed. 'Nah, it was a kind of early public relations exercise,' he said.

'It was?'

'Sure, if you wanted to grow a town, having it called "Murderer's Bar" wouldn't be such a good idea. That's what it used to be called, so they named it Happy Camp instead. It was better for business.'

Later still, someone told me that the town was named after James Camp, a man supposedly of cheerful disposition, who opened up a merchandising and drugstore in 1865. The rather grand brick building, with its corrugated tin roof and his name painted on the side, still stands today, one of a number of buildings that survive from the gold era. It's empty but it isn't going to fall down any time soon.

The official version sounded most likely to me, but even that could be challenged. One local member of the Karuk tribe, James A. Waddell, writing in *Happy Camp, CA: Bits of History* (2008), tells it slightly differently. In the story passed down through his family, the naming 'with a spot of whiskey' did indeed take place in July 1851, and he is able to put the names of Captains Gwin Thompkins and Charles McDermitt among the merry group.

However, the inspiration for the name may have come not from their simple elation but from prospectors who had found gold there a whole year earlier. According to Waddell, these earlier prospectors enjoyed their success and left before winter closed in, but not before nailing a sign to a tree near the confluence of Indian Creek and the Klamath River that read 'This was a happy camp'.

According to Waddell, the name Murderer's Bar was not applied to the town until 1852 after upwards of forty Karuk were murdered in cold blood by a group of miners. 'That gave the new town of Happy Camp the nickname of "Murderer's Bar",' he wrote. 'Other daylight raids occurred on the Klamath River, Scott River, and Salmon River in these early years of 1850s.'

I wanted to know more about the effects of this influx of miners on the indigenous population and so several times I went to the little Karuk Tribe Museum and asked to speak to the best historian for the job.

'You want to know about the *genocide*?' asked the lady on reception.

I was promised that someone would call me but no one did. I called into the museum a few days later to ask for some kind of official tribal history lesson – I felt I badly needed one – but again no one was available to see me. Instead, I was told I would probably have to make a hundred-mile round trip to meet the person who dealt with the media, but I was never invited to speak to him. I rang a few times more, wanting to discover the truth about Murderer's Bar and the treatment meted out to the Karuk, but it seemed that my presence was not welcome, at least not officially.

Unofficially, I was treated with kindness and respect by individual members of the Native American community, whose names came to surprise me. In histories of the town wonderfully preserved by the Siskiyou County Historical Society in *The Siskiyou Pioneer* there were families called Johnson, Guy, Humphreys, Anderson, Fisher, Webb, Effman, Frazier, Brown and a host of other names with European derivations unexpectedly preceded by Frank, Bob, Ellie, Tom, Jack, Kate, Bill, Hazel and so on.

Should I have been surprised? I don't know, but I had met Native Americans in the northern Midwest and many of those had hung on to their arguably more traditional names. The provenance of the names in Happy Camp told a story of its own, one explained to me by Hazel Davis Gendron, a locally treasured writer and historian, and Ellen Johnson, the receptionist at the Karuk tribal offices who took pity on me when no one else seemed keen to help. She took me to her lovely home to meet her husband, Bill. Hazel and Ellen were part Karuk; Bill wasn't but was a keen advocate of their culture.

Ellen, who was aged sixty-seven, explained that European names were taken by young Karuk women when they were bought by lonely miners. For example, her great-grandfather, Jack Buzz – whom I took to be Italian – gave twenty-four red woodpecker heads and an old horse for her great-grandmother, whose name is recorded as Kate (red woodpecker heads were highly prized).

Ellen said she believed that many of these transactions were entered into freely and willingly but Hazel, a charming eighty-five-year-old who had been recording Karuk history for more than half a century, told me that some of them were the result of greed and violence. She spoke of one massacre at Clear Creek in modern-day Siskiyou County in the 1860s in which large numbers of Native American men were wiped out. Their wives were then divided up among the white settlers.

'I believe my great-great-grandmother, known as Elizabeth, born at the Karuk village near Happy Camp, was probably one of those young women of the village taken captive when most of the men were killed or driven into the mountains during a massacre in the 1860s,' said Hazel. 'The miners chose from the women and many "half-breed" children were born. My great-grandmother, Emily, was probably among those, and born someplace on the coast, fathered by a white man. She believed his name to be Olmsdale or something that sounded like that.

'At any rate, when she was about twelve years of age, Emily and her mother made it back home to Clear Creek, and lived out their lives back in the area. Her mother did unite later with a farmer from Virginia whose name was John Henry Haley, a black man, and had more children by him prior to her death sometime before 1900. Emily went on to unite with a miner-rancher near Happy Camp and left many descendants in the area.'

Before the miners came, the Karuk had enjoyed something of an idyllic existence. They lived in small villages of semi-submerged cedar buildings, each with a sweathouse for the men and a menstrual hut for the women. They were a peaceful people with a quiet way of life. They had a special relationship with their surroundings, not least the mountains; the physical limitations placed on the Karuk by the Siskiyou and Marble mountains, the Salmon-Trinity Alps and the Red Buttes Wilderness gave them the impression that they were living at the centre of the universe. In fact, they had a name for the centre of their world: it was (and still is) a stretch of white water at Katimin called Ishi Pishi Falls on the Klamath, north of the Salmon River town today called Somes Bar.

They lived a life of plenty in the spring, summer and autumn, enjoying the bounty of the river, its salmon and steelhead or rainbow trout, crayfish and eels. They gathered acorns from the tan-bark oak to grind into flour for the winter and they hunted with bows, bagging duck, grouse, quail, elk and, most prized of all, deer. They believed that a deer, once eaten, was reborn. During the winter, they ate cured salmon and smoked meats while the brave young men of the villages would creep into caves in search of hibernating bears. The men would call for the bears to chase them, and if they did not the animals would be hauled out to their deaths.

From the profusion available to them the tribespeople gathered pine and hazel nuts, juniper berries, salt, bulbs, herbs, greens and a huge variety of plants that were either eaten or used for medicinal purposes. They traded with the Yurok for seafood to eat and for seashells that were valued in ceremonial dress and as adornments and money, the most valuable of all being dentalium shells, which are shaped like small tusks. They grew just one crop, tobacco.

The Karuk wore buckskin clothing and furs from squirrels, raccoons and other small mammals. Their basket-weaving – which was of exceptional quality – was prized by other tribes. Atop a Karuk head would normally be found a finely woven and brightly coloured cap. They enjoyed gambling for shells and feathers, valued the art of storytelling and made their own music for sacred ceremonies. For tools they used wood and stone, principally obsidian and soapstone.

'They were a people who respected the Great Creator in their daily living, believing that the well-being of the tribe depended on the prayers of the medicine man,' Hazel told me. 'They believed that the priest held communion with the Great Spirit and made good medicine for the Yurok and the Karuk, the downriver and upriver people. The prayers were to insure the health and well-being of the people, with ample game, fish and acorns for the year. Although some say that there was a chief in the tribe, the medicine man and his role seems to have been the most important within the Karuk tribe.'

The Karuk had their own moral code and justice system. When crimes or slights were committed, retaliation was not the first response – retribution was, in the form of compensation. Retaliation only came into play if compensation was not offered. Every crime had a price: for killing a man, the compensation for the bereaved would be fifteen strings of dentalium shells.

Imagine the effect on such a society, then, of hundreds and then thousands of armed men arriving, gold-hungry men who demonstrated no respect for your ceremonial or hunting grounds, who despoiled your rivers, ate your fish, depleted your sources of other food and, on occasion, shot your husband or raped your wife.

The anthropologist Maureen Bell, in her wonderfully informative book *Karuk: The Upriver People* (1991), says that the earliest record of white men, trappers, passing through Siskiyou County was in 1827. One oft-quoted and celebrated Karuk called Indian Ned remembers seeing white men for the first time in 1843. In 1849 a Major Pearson B. Reading crossed the coastal mountain range to the Trinity River and found significant quantities of gold, a discovery that found its way into the Sacramento *Placer Times*.

'By June 1850, a party of miners reached *Athithufvuunupma* (Happy Camp) for the first time, where they claimed that the Indians were so hostile they had to turn back,' wrote Bell. 'One year later a mining camp was established there, in July 1851 . . . Hundreds more soon followed. The Karuk greatly outnumbered the newcomers at first and were gracious and kind to them. They were helpful as guides, in crossing streams and in myriad other ways, for the Karuk respected these people as they did all others.'

Soon, however, the trickle of miners turned into a torrent and in 1852 more than 2,200 flooded into Siskiyou County. Thereafter, skirmishes between miners and Native Americans became commonplace.

'The worst, the most unscrupulous of American immigrant

society was let loose upon northern California,' wrote Bell. 'In April 1852, Redick McKee [a US Indian Affairs agent tasked with negotiating with Native American tribes] wrote to the governor of California to inform him of the Indian–miner conflict. In one incident, he relates that thirty or forty Karuk were murdered "almost in cold blood" by miners from Happy Camp.'

This was almost certainly the incident at what became known as Murderer's Bar.

Bell continued: 'By the end of 1851, according to the population estimates of [twentieth-century University of California Professor Sherburne] Cook, over half of the Karuk tribe had died off. The Karuk community had been devastated by disease, especially syphilis. The rivers had been contaminated with muddy water washed from the mines, and the Karuk, no longer able to catch the same amount of fish as in previous years, were starving.'

To the miners, the river meant gold, not fish, and they were not prepared to give it up. The Karuk became less trusting, particularly when food became more and more scarce and they were forced to watch their children starve. This went on for three decades, with levels of conflict rising and falling from time to time in what the white settlers came to call 'the Indian problem', the problem being that the Native Americans simply existed.

'The [white] Americans, whether intentionally or not, propagated genocide [against] the Karuk,' wrote Bell. 'They raped their women, infected them with unknown diseases, destroyed the habitats where fish and game had once been plentiful and outright murdered hundreds of individuals. The Karuk tried to defend themselves but they had neither the power nor the means to drive all of these newcomers away.'

It is estimated that there were around 2,700 Karuk in 1848. By 1910 the number had fallen to around 770.

Tragedies such as this were played out all over California during the Gold Rush. As well as syphilis and starvation, smallpox and influenza wreaked havoc among Native American communities, killing many thousands. I found it hard to believe the official California government figure of only 4,500 murders of Native Americans during the Gold Rush, but there it is. Whatever the figure, it would have been lower had the Californian government not put prices on the scalps of American Indians – twenty-five cents in 1856, rising to five dollars by 1860. This was an attempt at racial eradication as cynical as any in history. Many white settlers simply believed that life would be better without the indigenous population and so they enacted their own version of ethnic cleansing.

California governor Peter H. Burnett summed up the feeling among white settlers when, in January 1851, he said: 'A war of extermination will continue to be waged between the two races until the Indian race becomes extinct.'

According to Professor Cook, the Native American population in California fell from 150,000 to just fifty thousand between 1848, when Marshall found his gold, and 1855, when the Gold Rush petered out. By 1900 the population stood at around sixteen thousand.

11

Wild Bill, Duane Wilburn and the Rattlesnake

I was not avoiding Wild Bill, Duane and the rest of the miners at Independence Camp, but over the next few days I devoted time to prospecting on stretches of water twenty miles north of them. I lowered myself gingerly down a steep bank to a cool and quiet spot on Indian Creek where the original 'Happy Campers' had found gold, feeling more isolated here than anywhere I had worked so far. I found another location a short distance south on Elk Creek that was so vibrantly fertile – the air with spores, the water with fish and the soil with ferns and orchids – that I actually felt nervous, as if I were somehow interfering in the business of nature. My fear of bears and mountain lions had not abated, yet my trepidation brought with it a heightening of the senses and a feeling of being so very alive, right here, right now. I continued to find small amounts of gold. Several times I called in on Wild Bill at Independence but he never seemed to be there.

I began heading much farther south, thirty, forty miles, to the North Fork of the Salmon River, where I prospected on gravel bars at wider stretches with white water and clear, fresh air. I found a few flakes of gold on the Salmon River, and each time colour revealed itself in my pan my heart leapt and I realized that this was not like other kinds of drug; you did not need more of it to achieve the same high. A tiny amount goes a long way, and this allows you to deceive yourself into thinking you have a lot when you have almost nothing at all.

At night, I told Tom, Craig and Terry how I was doing and they might have been puzzled by the odd juxtaposition of my lack of success and my satisfaction with it. I had hardly any weight, but I had colour and, for now, that was enough for me. I was beguiled by it and several times asked myself which I would prefer, an ugly and bulbous two-ounce nugget in my hand, or one ounce beaten flat to a surface area of a hundred square feet. The answer was always the same: a hundred square feet. Did that mean I did not have gold fever, as I would happily have forsaken two ounces for one? Or did it mean I had become bewitched by the beauty of it and blinded to its true value? And what was its true value anyway if a man might achieve happiness by moving one-fifth of a ton of earth and finding one flake worth just a few pennies?

Disappointed by how paltry my gold looked in its vial, I filled the vial with water. Refraction acted on its contents and I could see each flake much more clearly. I never took the vial out of my pocket now. I didn't want Marcy to see it.

Terry and JoAnne took me out with them, heading south down the Klamath to a steep bank that we negotiated with the use of a rope. I really must get some, I thought. Terry had tied up a small dinghy and, after hauling our equipment into it, we headed downstream on a slow diagonal to the east bank. I felt grounded when I was with Terry and JoAnne; she saw through my bluster, while he ignored it.

We disembarked and moored under a thick grove of willows where, tied up from the day before, was Terry's dredge. We would not be dredging today, he said; we would be high banking. It was 110 degrees.

The couple had been working this spot with some modest success for several weeks. I was aware that to be taken there for a share of that, even for one day, was a privilege and I decided to show my gratitude by working furiously. Aside from the dredge, they had pumps and plastic hoses down to the river and a settling pond, a large vacuum hose powered by a motor, a sluice positioned on legs at belly height and a canvas gazebo where JoAnne positioned herself to begin digging. All this equipment had been left, in the certain knowledge that no one would steal it. Theft was considered anathema among miners.

I soon persuaded JoAnne to let me do the digging and I set about it with a gusto that worried her.

'Take it easy,' she said. 'If there's gold there, it's not going anywhere.'

But I ignored her. Sweat poured down my face and when Terry came up from the river – he had decided to do a little dredging after all – he told me to slow down.

'You peed yet?' he asked after a couple of hours.

'No,' I said.

'You're dehydrated. Drink some water.'

I drank deeply and quickly. There wasn't a moment to spare.

I was digging closer and closer to bedrock when a large rock stunted my progress. I worked around it with my pick until it loosened and came away from the packed earth. I lifted it up and gasped. It was covered in flakes of gold. I checked one or two to see whether they were iron pyrites but they weren't. Carefully, so as not to dislodge any of the precious metal, I climbed out of my hole and carried the rock down to the river bank where JoAnne and Terry were working in the shade.

'Looks promising,' said Terry.

'Not bad,' said JoAnne. Neither seemed very impressed, but inside I was ecstatic. I washed the rock in a bucket of concentrates in the sure knowledge that I would see the gold again once we set to panning.

We broke for lunch and I asked Terry what had made him choose this location to dig. He took a swig of water, stood up and told me to follow him. We walked farther downstream and he took me over a low bank and into a wide clearing with grass-covered hills on three sides.

'That's why,' he said.

I looked around but could see no clues in the terrain.

'The hills,' he said. 'They're man-made. The first 49ers came and picked up the surface gold – the placer gold – using pans and rockers and sluice boxes. But after a while there was nothing left that could be found that easily, so they began using high-pressure hoses that washed away entire hillsides and mountains, and then they processed the loosened earth. It was called hydraulicking.'

Miners would dig deep ditches to capture water high on the mountainsides. This was then diverted down wooden flumes and into first hoses and later huge iron water cannon that spewed out jets at pressures of five thousand pounds per square inch. This washed away tens of feet of topsoil and exposed ancient riverbeds and gold-bearing gravel. The foothills of the Sierra Nevada were devastated, and rivers were blocked and diverted by the debris that this process generated, leaving behind a landscape that might have been shaped by a nuclear explosion. Downstream, farms died from thirst. Rivers became impassable.

Bayard Taylor, the American poet and travel writer, visited the region after years of hydraulicking had taken their toll and was shocked by what he found. He wrote: 'Nature here reminds one of a princess fallen into the hands of robbers who cut off her fingers for the jewels she wears.'

Hydraulicking – in fact, all mining activity that could adversely affect agriculture or the safe navigation of rivers – was finally outlawed in 1884. If you look carefully, you can still see evidence of these processes everywhere in strange hillocks, gullies and misshapen mountains, but for the most part nature has made it beautifully invisible to the untrained eye.

Terry's experience had told him that such large-scale operations often overlooked gold at the small scale in which we were interested, so it might still be lying there for us to find. He was a canny operator. At the end of the day we ran six buckets of concentrates through the sluice and gathered half a bucket of pay dirt that we would pan back at Big Foot. I helped the couple to load up the equipment that could not be left behind, thanked them and headed off to Independence where, finally, I found Wild Bill. He was going to take me out prospecting at his golden spot, I just knew it.

'Hey,' he said, deriving no particular pleasure from seeing me. He seemed distracted, watching one of the osprey wheel overhead from the corner of his eye.

'Got your note,' I said.

'I got good news for you. I spoke to Duane and he says he'll take you out with him. Gotta be here by six tomorrow or he'll go without you, okay?'

I thanked him with feelings of disappointment and trepidation. Covered in mud, sweat and mosquito bites, aching to the point where my eyelids hurt, I turned tail as darkness fell and headed back to camp.

I slept for ten hours and woke with cold and heavy limbs and a sense of dread. Would I be able to put in another day's work like yesterday? Would I fall under Duane's equipment, leaving him casually to tell the other miners how I had screamed pathetically and wet myself before dying quickly and without a fight?

I stocked up on water, bread and cheese and headed south. I was running late. This was out of character and I wondered whether subconsciously I was trying to miss my appointment with Duane. People here didn't suffer fools gladly and I was coming to the conclusion that I was a fool. So far, I had successfully hidden this fact but it would be only a matter of time before I would be found out. Of the miners I had encountered, most prospected for fun but some, Duane included, tried to make a living out of it and took their work very seriously. It would not be long before he would come to despise me for the tourist I really was.

When I arrived at Independence, Duane was already up and had some coffee brewing. The embers of a fire lay in a pit outside the small trailer he called home, and his dogs rose up from it and began to sniff my trousers.

'Mornin',' he said, nice as pie, and we shook hands. 'Get you some coffee?' He was wearing combat camouflage pants and hat, with a green T-shirt, and the first thing I noticed was that his white goatee was much longer than I remembered.

He poured me a strong black coffee and we both watched as a four-wheel drive turned off the road and on to the dusty flat. The driver climbed out, breathed in the morning air and, looking down the bank to the white beauty of the Klamath, nodded a hello. I nodded back. Duane didn't know this man, but within minutes they had pulled out their rifles and were comparing the range or accuracy of one against the other. They would tug, slide and click parts of each other's weapons, pulling this apart, staring down that and nodding or suggesting a limitation here, a plus-point there. Then they got out their handguns and began talking about weight and balance and what they had blown away with this or that.

'No one can boss you around when you got one of these,'

Duane said to me. I was out of my depth and suppressed an overwhelming urge to laugh hysterically, as if this would somehow fill the void in my knowledge of firearms.

Except, I thought, *someone with two of those.*

The driver of the four-by-four was called Brian, a sixty-four-year-old former forestry engineer from Gig Harbor, Washington state. He wore braces on his jeans, a grey T-shirt and was shaven-headed, preferring instead to wear his hair on his chin in a long white beard. 'Here,' he said to me. 'See how this feels.'

He gave me a pistol and I did what ninety-five per cent of Englishmen would have done in this situation: I struck a pose like James Bond. As I moved the gun through the air in order to bring it coolly to rest across my chest, both men ducked and yelled 'Whoa!', as inadvertently I pointed the barrel at their respective heads.

'You crazy?' asked Brian, relieving me of the weapon. I took this as proof that an idiot with a gun is much more frightening than a sensible person with one. In most situations you can scare the living daylights out of bigger, better-armed individuals by doing something very stupid indeed. Often, just being English is enough.

I once walked out of a courtyard filled with heavily armed men in Baghdad – men who I thought might kidnap me – by behaving like a person who was so dim that he did not know he was in mortal danger. Having stumbled into the situation, I shouted 'Hello chaps!' and approached each one loudly and with my hand outstretched, respectfully greeting them with the words 'Salaam alaikum', which means 'Peace be with you'. Instinctively, these utterly bewildered gunmen were so polite that they felt obliged to tuck their AK47s between their knees so they could return the greeting and shake my hand. Then I left.

Of course, the kidnap threat might not have been real; I will never know. But I do allow myself to imagine them arguing after my departure.

'Why did you not take him?'

'Why did *you* not take him?'

'Where is the challenge in taking someone so stupid?'

'No one would pay the ransom.'

'Well, I liked him.'

'He *was* nice, wasn't he?'

'Unbelievably stupid, though.'

When Duane and Brian had finished comparing their weapons they said goodbye and off we went to Duane's secret spot on the river. We took my vehicle as his was out of gas. He was intense and refreshingly honest, and as we drove along I realized that he was trying to put me at my ease. This left me wondering whether he knew he could make people uncomfortable with his wild appearance and intensity, and I came to the conclusion that I simply didn't know and had almost no intention of asking.

'Can I smoke in the car?' he asked.

'You can do anything you want,' I said.

'Great.' Duane thought about resting his boots on the dashboard but something told me he was on his best behaviour. Instead, he lowered the window and blew out a cloud of blue smoke. 'You got gold fever, huh?'

'No,' I lied. 'At least, I don't think so. Have you?'

'Well, sure. Wouldn't be here if I didn't. What's wrong with gold fever? I *love* this way of life and I wouldn't be living it if it wasn't for the gold. Look at it here . . .' He let his right arm surf the air outside the window. 'It's goddamn beautiful. Imagine all the poor bastards working at a desk in some office or in a factory, watching the hands on the clock go round. What time is it here? Hell, *I don't know!*'

There wasn't an ounce of fat on Duane and the veins in his arms and legs popped out of his skin, forced to the surface by muscle vying for space in his hard, crowded body. He was used to unrelenting work and I had the feeling that if a pump or winch or piece of equipment that might save your life broke, Duane could fix it. He seemed like the kind of man who understood torque and horsepower and grades of oil. We stopped, unloaded our gear, and he lowered two long lengths of rope, tied in the middle, down a steep bank to a depth of about sixty feet. The incline was at fifty degrees. This was the bank down which he had fallen, his trolley full of equipment careening across his middle. He seemed anxious that I shouldn't suffer a similar fate and demonstrated an unexpected concern for my well-being. 'There,' he would shout, ' left, left, put your foot on that rock, down a little, now right. Watch out for that bush – it's poison oak.'

I suddenly realized that if I were going to do something as spectacularly out of character as ford a river, run the gauntlet of bears, rattlesnakes and mountain lions, dig and pan for gold, then it was possibly Duane in whose company I would most like to be uncharacteristic.

He had a grey Mariner dinghy hidden in the undergrowth at the foot of the ravine and on it was a solar panel hooked up to a twelve-volt battery that powered a motor. 'Swapped it for a sluice box,' he said, smiling. 'She gives me thirty-five pounds of thrust – not enough to push against rapids, but enough to get me from one side of the river to the other.'

He steered the boat carefully so as to use the current to carry us across the Klamath, and then he found a slow-moving stretch up which the motor could crawl, travelling against the flow of water for the last twenty yards to his landing point.

We unloaded our equipment on to packed mud among willows and what appeared to be young birch trees. The bank rose steeply to a height made uncertain by thick fir; everywhere there were dark granite rocks that had been carried here by prehistoric floods, and smoothed and rounded by gentler flows. We walked about two hundred yards upriver, climbing over boulders and through river grasses until we came to Duane's diggings. It was clear why we were here. The rocks, perhaps five or ten feet above the summer water-line, would be under water during the spring melts and they would slow down water and capture any gold that the river was carrying.

Duane had already been working one area and he picked out another for me. 'Here,' he said. 'Reckon this is a good spot. You got one of these?'

He threw me a trowel but I said I already had one.

'This is better,' he said. 'You keep it.' Mine was made from plastic compound; the one he threw was steel and thinner, the better for scraping into crevices.

'Thank you,' I said.

This was the first of many kindnesses shown to me by Duane Wilburn, and I began to find him fascinating. Once I got to know him a little better, I refused to believe his claim to be aged fifty. His face was lined and leathery from the sun, but he had the body of a twenty-five-year-old. He was a carpenter by trade, originally from Tacoma in Washington State, and he had served in the US Army from 1981 to 1993, with several tours abroad, including Panama, Germany and South Korea.

After leaving the military, Duane had set up his own car-pentry business but it had collapsed when he was diagnosed with cancer: seventy-five per cent of his stomach and twelve feet of intestines had had to be removed. He had undergone chemo and radiotherapy, during which time his business had fallen apart. Eventually, he beat the disease.

'I was so sick that I lost my company but social security wouldn't help me,' he said. 'They said I wasn't entitled to anything even though I had served my country.'

With the last of his money he'd bought a travel trailer and headed off into the wilderness, camping, fishing and bow hunting for survival. Two years earlier, he had stumbled into Happy Camp and learned about prospecting, gradually building up his equipment as and when he found colour in his pan. When I met him, his stock among other miners was high. He was a man who knew what he was doing, they said – and there was no greater compliment than that.

'I don't have much faith in authority,' he said. We had stopped for a break in the shade as the temperature hit 113 degrees. 'Men and women in authority just want to further their own wants and needs. They don't really care about the people. I don't have any need for authority or government here. We look after ourselves and the people are real. There are no white lies or any of the grey areas in between. It's a more honest way of life. I don't want to live around people who can't be truthful.

'When I first came here, I knew nothing about prospecting or claims or such things. I had two inner tubes from a truck and I used to jump into the river with them, one carrying my equipment, one carrying me with a bottle of vodka and a cocktail shaker. I used to float down the river and look for gold wherever the current took me. I don't do that any more because now I know you gotta be respectful of other people's claims.

'I try to keep it simple and when I prospect I prefer digging and panning to dredging or high-banking. Panning is purer, truer mining and I think God looks more kindly on you if you just go about your work quietly. When you're panning in the river with the birds flying overhead, it's soothing and humbling. You feel as if you're close to nature. You can find perfect solitude if you look for it. Sometimes it finds you.'

In a good week, Duane expected to find about five hundred dollars' worth of gold; in an average week maybe three or four hundred. Even less than that was enough to support his lifestyle.

'I have no bills,' he told me, rolling a cigarette and swigging on some water. 'I have to pay insurance for my vehicle, I have gas and food for me and the dogs, but that's about it. If I need something big – like the motor for the boat – I can usually work out some kind of trade. I have no worries, only excitement each time I find gold.'

I have often met people living alternative lifestyles and found that if you probed hard enough, a sense of longing would emerge. Their troglodyte existence in caves warmed by hot springs was free and eco-friendly and wonderfully communal, they might say. But then the feeling surfaced that they would trade their youngest child for access to a window and a breath of fresh air.

I had no such feeling with Duane. I had the impression he had experienced a lot of pain, and he told me about failed relationships and three children in their twenties whom he hardly saw. Instead, he had the demeanour of an injured person whose wounds had been draped in a most effective salve or balm. He seemed relieved, as if finally breathing out after years of breathing in. If there was a God, and if he really did look more kindly on Duane's peaceful and harmonious way of mining, then perhaps this was Duane's true reward and the gold was merely a by-product. I wondered whether health might not be his prize too, one infinitesimally lighter but much more precious. Duane told me that his physical pain had gone, his digestion – in a depleted digestive system – was better and he felt as fit as he had when he was thirty.

'I have no stress,' he said, 'and I breathe clean air, eat clean food and drink clear water. I have never felt healthier.' Later he would take me to a small creek fed by a spring from which he sourced all his water, and I would wince inwardly and feel obliged to drink from it, wondering what protozoa might be entering my system, what dead animal was rotting farther upstream. It was the sweetest, purest water I had ever tasted and from time to time I would return for more.

At the start of the day I had asked Duane what was the polite way of mining; would we share everything we found that day or would it be every man for himself? In common with all the small-scale prospectors I met, he opted for the latter. Where there were individuals digging individual holes, miners preferred to trust in their own luck; where there was a bigger enterprise and a process that involved one person doing one job, another doing something else, and another doing something else still, then those proceeds would be shared out. I say this because during the morning I felt I was doing rather better than Duane and felt guilty enjoying such success on his claim.

After lunch, however, I moved to a new location a small distance away and my finds reduced to nothing. Duane, on the other hand, did quite well, and I was pleased to see his good fortune. I was washing another colourless pan at the river's edge when I suddenly heard the sound of maracas behind me and my mind skipped back twenty years to a bar in Mexico, to endless margaritas and a beautiful girl who kept calling me Esteban . . .

I was woken from my daydream by the jarring sound of steel on rock. Duane was yelling, '*Rattler. . !*

I leapt up to see the blade from his shovel chopping clean through the back of the snake's head. It had been right behind me, and even though it might have given me a fatal bite, I felt sorry for it. It was in retreat when Duane killed it and later, dry-mouthed, I asked whether he might not have left it alone.

'You can't leave a rattler alive in a spot where you're prospecting,' he said, skinning the corpse. 'You gotta move rocks and they like to cool under rocks and if you don't kill him now then you're gonna meet him again and next time he might kill you.'

Of course, I knew Duane was right and I told him I was grateful. He had taken off his shirt and I looked at the tattoos that covered his body. There was the cobra, his first, drawn on his eighteenth birthday while he was in the Marines; the scorpion from Thailand; the rose with thistles in honour of his mother; there were Pegasus and the unicorn nodding to a Celtic heritage. Randomly, there were self-drawn tattoos of magic mushrooms, a dragon, a banshee, red arrows and wings, and there was the symbol for the chemical compound THC, the active ingredient in marijuana.

I felt ashamed for judging Duane before taking the trouble to understand him. I visited him again and came to value his company and his advice. Would it be stretching the truth to say that he saved my life? Probably. Would he have stepped up to save me in other, less ambiguous and more dangerous circumstances? Yes, with all my heart I believe he would.

12

The Point of No Return

In *Eldorado: Adventures in the Path of Empire,* published in 1868, the writer Bayard Taylor described the hardships that faced the overland Argonauts beyond South Pass, the half-way point of their journey. He wrote:

> Here the pasturage became scarce and the companies were obliged to take separate trails in order to find sufficient grass for their teams. Many, who, in their anxiety to get forward with speed, had thrown away a great part of the supplies that encumbered them, now began to want and were frequently reduced, in their necessity to make use of their mules and horses for food.
>
> It was not unusual for a mess, by way of variety to the tough mule meat, to kill a quantity of rattlesnakes, with which the mountains abounded, and have a dish of them fried for supper. The distress of many of the emigrants might have been entirely avoided had they possessed any

correct idea, at the outset of the journey, of its length and privations.

William Swain's route with the Wolverines took him four hundred miles over mountain ranges and rivers running north–south that repeatedly halted their progress. The original route of the California Trail led north to Fort Hall, then south-west along the Snake and Raft rivers in present-day Idaho. But recently, some Argonauts had left this trail near Soda Springs, Idaho, and blazed a new one heading due west. It was called Hudspeth's Cut-off and it was utterly uncharted; this new route was taken by the Wolverines on 15 August, spurred on by false rumours that it would get them to California faster. This 'short cut' rejoined the old California Trail about sixty miles south of the confluence of the Snake and Raft rivers at Cassia Creek; it actually saved William and his comrades only a few miles.

One-third of the overlanders, Sarah included, went via Salt Lake City, Utah, farther to the south, and most rejoined the California Trail at City of Rocks, a haunting volcanic array in Idaho, 180 miles north-west of the Mormon city. From there, the route led towards the Humboldt River, which slices south-west through Nevada and into the Great Basin, a vast and arid region comprising mostly desert.

Possibly because of Sarah and Josiah's lateness – they did not leave Salt Lake until 30 August – they did not head north to rejoin the trail. Instead, they headed due west across the treacherous Great Salt Desert. The only company they had was an ageing man they had taken along in return for his ox, though later they would pick up two more stragglers; they were probably the very last party to attempt the journey along this route in 1849. Some of the Mormon elders at Salt Lake pleaded with them to remain there for the winter; what grass existed in the Great Basin would have been either eaten by the thousands in front of them or scorched by the summer sun. And if their animals expired, so would they.

Their pleas fell on deaf ears. Sarah wrote: 'They told us we would lose our cattle and perish on the desert; or, if we reached the Sierras, would be snowed in and perish there. We heard it, we coolly talked it over and yet, so perverse were we, that on the 30th day of August, a solitary wagon, drawn by three yoke of oxen, and in charge of only two men, left Salt Lake City, bearing as its passengers one woman and one little child, and for freight only so much provisions as might last us till we could scale the great Sierras and reach their western foot.'

Crossing Nevada's Great Basin would have been impossible without the Humboldt River. In places more like a stream only a couple of feet deep, it flowed through 350 miles of desert before disappearing, not into the sea, but into the ground at an area appropriately called the Humboldt Sink. It sustained the lives of the oxen that pulled the Argonauts towards their destination and provided the parched grass and wild sage that would maintain their strength but, in truth, the overlanders loathed it for its dank saltiness and the stagnant pools and depressing landscapes that attended it.

One wrote of it: '[The] Humboldt is not good for man nor beast .. and there is not timber enough in three hundred miles of its desolate valley to make a snuff-box, or sufficient vegetation along its banks to shade a rabbit, while its waters contain the alkali to make soap for a nation.'

Most of the Argonauts travelled at night to avoid the withering heat of the day. They were weak and tired and complained of clouds of sand and dirt that blinded them and made them constantly thirsty. On 6 September, William Swain wrote: 'We made a start at [7pm] and travelled along in dust, or rather volcanic ashes, from six to twelve inches deep and light as mist filling the air with clouds which enveloped the teams, wagons and men, along the whole of this valley, and we are getting disgusted with this dirty, dusty, tedious killing of time. We find none but dry feed, and that has been used up by the emigration preceding us.'

The three-week journey down the valley of the Humboldt was hard and depressing. Food supplies for most ran low. Nothing that grew there could be eaten by humans and there was precious little game. Diarists spoke of constant attacks by groups of Native Americans who they disparagingly called 'Diggers' for their propensity to scavenge in the ground for food – small mammals, seeds and insects. At night, the Diggers would fire arrows into oxen and then disappear, returning the next day to feast on the abandoned animals. Their carcasses lined the route, as did the heaps of possessions left behind by Argonauts with too few animals to haul their wagons.

The Humboldt provided a lifeline for the Argonauts but not one that would see them safely to their destination, for the sink – the place where it vanished into unforgiving sand – left them forty miles short of the Carson or Truckee rivers, their next sources of water, on the other side of the aptly named Forty Mile Desert.

Astoundingly, Sarah and Josiah would attempt their journey from the Mormon city with no more intelligence than some handwritten directions scrawled down by a stranger. 'Our only guide from Salt Lake City consisted of two sheets of note paper, sewed together and bearing on the outside in writing the title "Best Guide to the Gold Mines, 816 Miles, by Ira J. Willes, GSL City",' Sarah wrote. 'This little pamphlet was wholly in writing, there being at that time no printing press at Salt Lake. It was gotten up by a man who had been to California and back the preceding year. The directions, and the descriptions of camping places, together with the distances, seemed pretty definite and satisfactory until they reached the lower part of the . . . Humboldt River; when poor camping and scarcity of water were mentioned with discouraging frequency. From the sink of the Humboldt, all seemed confusion.'

The man who sold them the 'pamphlet' said that they would bump into a group of Mormons returning from California. These good folk would give them directions to the best places to camp, where there would be water and grass for their oxen. The word 'haphazard' doesn't come close.

After hearing horrible stories of privation and death in Forty Mile Desert – precisely the route Sarah and Josiah had chosen – the Wolverine Rangers decided again to take a new route, called Lassen's Cut-off, in order to miss the Humboldt Sink and the desert altogether. Established by Peter Lassen, a Danish-American rancher of dubious character (he advertised the route as being much shorter than it actually was, taking travellers on to his property, Lassen's Ranch, in the northern Sacramento Valley, where he sold them provisions at vastly inflated prices), this cut away from the Humboldt ninety miles north of the sink and took them in a westerly direction towards the Sierra, but it had dangers of its own, not least a seventy-mile stretch where there was no feed at all for the oxen. The fifty-six men of the Wolverine Rangers gathered as much grass as they could to sustain their animals and set off along the cut-off on 21 September.

Imagine William's disappointment when he realized that far from being a quicker, safer route to the Sierra Nevada mountain range, beyond which were the goldfields, the Lassen Cut-off was more arduous and longer by two hundred miles than the route they had avoided; it drove them through canyons and across deserts that steered them not due west, but ever farther north towards Oregon.

They passed through the Antelope Hills in present-day Nevada and within days found themselves in Black Rock Desert, exactly the kind of environment they had hoped to avoid. Here, they were surrounded by the rotting carcasses of fallen cattle, the possessions of those who could no longer carry them and by the graves of Argonauts whose journeys had come to an end. At night, temperatures plummeted below freezing; during the day they rose above a hundred degrees.

Before Black Rock Desert the Wolverine Rangers had been told by other ill-informed travellers that they could stock up on water at a place called Rabbit Hole Springs, but it was not the kind of oasis they had imagined. On entering the desert at night, William noted: 'a destruction of property beyond my conception lined the road. Wagons and carts were scattered on all sides and the stench of dead and decaying cattle actually rendered the air sickening. Some idea can be drawn from the fact that in one spot could be seen one hundred and fifty dead creatures . . . Many of our cattle tired on the road and were left to perish by hunger and thirst.'

Conditions improved for a while and the company found good grass and water but by then they were breaking through ice each morning to drink. By mid-October they were in the mountains proper, but the wrong ones. They climbed upwards through mountain pass after mountain pass until they reached what they thought was the summit of the Sierra Nevada. In fact, they were crossing what are today called the Warner Mountains; the Wolverines were actually in California, but far north of their intended destination.

On 13 October the Wolverines met a group of soldiers who had been sent out along the various gold routes with emergency supplies to help straggling and desperate Argonauts reach their destinations before winter set in. The Wolverines were only one day ahead of the last party on the Lassen route. The soldiers gave them beef and the bad news that they were still almost 400 miles short of Sutter's Fort and 220 miles from Lassen's Ranch. On 18 October William spied Mount Shasta and was probably as awestruck by it as I had been. His party struck out south down the Pitt River and Feather River Valley, battling snow and blizzards, fording icy rivers and crossing seemingly impassable routes.

He arrived at Lassen's Ranch on 8 November. His journey had lasted 211 days.

Describing the desperation of the stragglers – who by then were disproportionately made up of slow-moving families – William wrote: 'Had manhood in its strength been doomed to surmount these dangers alone, human suffering would have been less. But there were the infirm and aged, for even here were grey heads. Many of the emigrants were palsied by that terrible disease, scurvy.

'Here, too, were females and children of every age. Here might be seen a mother wading through the snow and in her arms an infant child closely and thickly wrapped with whatever would secure it from the storm, while the father was close at hand exerting to the utmost in getting along his team, wagon and provisions – the last hope of securing the life of the family. There might be seen a mother, a sister or a wife, winding along the mountain road, packing blankets and other articles, followed by children of every age, each with some article on his back or in his hands, hoping thus to enable the teams to get along by lessening the load.'

Haphazard their plans might have been, but by luck, inevitability or divine intervention, Sarah and her husband did, indeed, meet the Mormons that the purveyor of their raggedy map had predicted they would. These were the travellers who would advise the Royces on the best route to take across Forty

Mile Desert from the Humboldt Sink – and this they did to the best of their ability.

Using the sink as the starting point for the journey, the leader of the Mormon company drew a map in the sand. After the sink, they would see a wagon track leading off to the left that would take them to a grassy meadow, still green and lush, some two to three miles away, he said. They would find a well there, too, and should take the opportunity to rest for a few days, feed their animals and fill every vessel they had with water.

Further, Sarah wrote, he advised: 'Once out on the desert we were to stop at intervals of a few hours, feed some of the hay to the cattle, give them a moderate drink, let them breathe a short time and then go on. In this way, he said we would be able to reach Carson river in about twenty-four hours.'

If you have ever stopped a stranger to ask for directions, you might have some experience of what it is like to be disappointed by the accuracy of their information. You might also have come to a halt after becoming lost, to remonstrate with your partner – to gloat, perhaps – that you should have turned right back there, and not left. This might lead to a sullen journey but it rarely results in death. However, death was precisely what Sarah and Josiah were risking when they unquestioningly accepted this advice.

As it happened, the directions were correct but, travelling in the cool of darkness, Sarah's party missed the left fork in the road that the Mormon had suggested. Within a few hours the prospect of failure was very real and horribly terminal in a landscape littered with the bones and graves of other doomed crossings. When the fork failed to materialize, it began to dawn on the party that they must have passed it long ago; they were now many miles into the desert without the water or feed for the animals that they had expected to take on. They had only six pints of water for five adults, a baby and six oxen travelling in baking heat.

The party spent several wasteful hours trying to decide what to do next; their lives would depend on making the right choice. They decided to turn back, try to find the left fork and the meadow, rest for a few days and attempt the crossing again. It was 4 October.

'Turn back!' wrote Sarah. 'What a chill the words sent through one. *Turn back*, on a journey like that; in which every mile had been gained by most earnest labor, growing more and more intense, until, of late, it had seemed that the certainty of *advance* with every step, was all that made the next step possible.'

On the way back to the fork, they encountered a small caravan of wagons, the occupants of which had been camped the previous night at the very meadow the Royces had missed. It was about sixteen miles away, they said, which seemed an endless distance to the hungry, thirsty party and their animals. The caravan did not have enough provisions to take the Royce party along with them, but one of their number, a woman, spoke to Sarah and expressed concern for her safety and that of her child. They quickly went their separate ways but this woman, unknown, unnamed, would later save their lives.

Sarah's party rested and watered at the meadow and attempted the desert crossing again on 9 October, becoming increasingly horrified by the presence of large numbers of abandoned wagons and dead animals. 'We seemed to be but the last, little, feeble, struggling band at the rear of a routed army,' wrote Sarah. Among the debris, they found whole sides of bacon, trunks, clothing and furniture.

By morning the oxen had begun to raise their heads and quicken their pace; they could smell the water of the Carson River. Of course, beyond that they still had to cross the Sierra Nevada Mountains and it was now terribly late in the year, but as the mountains rose up to greet them, Sarah spied a trail of dust coming closer and closer. The trail was made by two mounted soldiers from the same relief effort that had aided William's party. They arrived with two mules, some food and priceless advice on how to cross the mountains swiftly before snowfall would make them impassable. The first piece of advice was to abandon the wagon and pack only as much as the mules could carry, leaving behind the rest of their possessions. Sarah was convinced that these men were Heaven-sent.

She wrote: 'As they came near they smiled and the forward one said [to Josiah], "Well sir, you are the man we are after . . . You and your wife, and that little girl, are what brought us as far as this. You see, we belong to the Relief Company sent out by order of the United States Government to help the late emigrants over the mountains. We were ordered only as far as Truckee Pass. When we got there we met a little company that had just got in. They'd been in a snow storm at the summit; most got froze to death themselves, lost some of their cattle, and just managed to get to where some of our men had fixed a relief camp."'

As soon became clear to Sarah, the unnamed woman they had met earlier had not forgotten them. The soldier continued, '"There was a woman and some children with them and that woman set right to work at us fellows to go on over the mountains after a family she'd said they'd met on the desert going back for grass and water cause they'd missed their way . . . and she knew they could never get through them canyons and over them ridges without help . . . You see, I've got a wife and little girl of my own; so I felt just how it was; and I got this man to come with me and here we are."'

Riding the mule with Mary in her arms, Sarah completed the journey over the highest peak by 19 October and then enjoyed a descent into the warmth and beauty of the Sacramento Valley.

'California, land of sunny skies – that was my first look into your smiling face,' she later wrote. 'I loved you from that moment, for you seemed to welcome me with [a] loving look into rest and safety. However brave a face I might have put on most of the time, I knew my coward heart was yearning all the while for a home-nest, and a welcome into it, and you seemed to promise me both.'

As I read of Sarah and William's journeys to the goldfields, the idea of one of my own began to take shape. I was still finding colour in the Klamath and its tributaries; some, like Dave Mack, were finding it in healthy amounts, but they were using the legally ambiguous form of dredging, and I had no dredge.

The previous night I had crept into my tent, zipped up the entrance and emptied the gold from my vial on to a plate, being careful not to lose a speck in the dank chaos that surrounded my sleeping bag. It shone brightly in the light of my lantern and seemed to cover almost a quarter of the plate's surface when carefully arranged. I sat, cross-legged, looking at it for a while; I'm not sure how long. I had begun to recognize the shapes of individual flakes, some like countries, others like flowers and silently I heard myself greeting them. I remembered one that was shaped something like a face but I couldn't find it. I used the blade of a knife to move the gold around and experienced mild panic when it failed to appear. Then, like a face upside-down, there it was. I sighed. Outside, a dog barked.

I estimated that I had hundreds of pieces of gold now, but each one of them was minuscule. They were thicker than gold leaf but not by much, and I realized that if they were beaten flat they would not cover a fraction of the surface area to which an ounce of gold could be pressed – one hundred square feet. At this thickness, they covered a quarter of a plate; beaten, maybe a few square feet. I was way, way short of even an ounce.

I had always considered myself lucky, but the gold nuggets and pickers I had expected to find had eluded me. So, too, had the pockets of dust, missed by generations of miners not as lucky as me. I doubted the gold I had found so far would cover the cost of food and gasoline.

I found I was becoming impatient with my inability to find gold in any significant quantity and that impatience was sometimes taken out on others. One morning, a miner approached me outside Big Foot as I was preparing for another day's prospecting.

'You hear about the guy on Indian Creek?' he asked.

The guy, I thought, always *the guy*. Every day someone would tell me about a big gold find, about *the guy* who found a nugget *this big*, and their hand would be shaped something like a walnut or a clementine.

'Heard about *the guy* on the Bear River . . .?'

'. . . the Yuba . . .'

'. . . the Feather . . .'

'. . . near Mariposa . . .'

'. . . Colfax . . .'

'. . . Grass Valley . . .'

'. . . pulling them out this big . . .'

'. . . this big . . .'

'. . . *this* big . . .'

'. . . swear to God, *the guy* is *shitting gold nuggets*.'

Who was he? If there was such a person, then surely it was Dave Mack. But even his luck didn't match *the guy*'s; *the guy*'s hauls were momentous. If you could find him, just following in his wake might make you rich. But he was as elusive as gold itself.

'Guy's pulling out nuggets *this big*,' said the miner, holding his fingers a plum's width apart. 'Swear to God. Heard he hit a real good pay streak and she just keeps giving. Goin' up there later to see what he got, maybe work upstream from him.'

I didn't know who this man was but I had seen his kind a hundred times before and was growing tired of him. Four-by-four, monster wheels, lumberjack shirt, baseball cap, Foster Grant sunglasses and a big mouth. He was only here because he didn't have the imagination to prospect anywhere on his own and that annoyed me. His innate insecurity rendered him incapable of doing anything alone. He had nothing to contribute and nothing new to convey and so passing on second-hand news – even when it was a lie – gave him, in his own mind at least, some kind of status. You would rarely see him prospecting solo; instead, he would piggyback on the hard work of others. I had done that for a time, in order to learn. But out of respect I had moved on and was now on a path of my own. I spat on the floor.

'Guy's pulling out *diamonds* at Somes Bar,' I said.

'Wow,' he said. He wasn't entirely sure about the veracity of this but he would be passing on the news first chance he got.

Cynicism and curiosity conveyed me to the spot on Indian Creek that he had described but there was no one there. There was evidence of some small diggings but nothing more than that. Much of the creek was shaded but here the sun shone, relentless and hot. I put down my gear and waded in, boots and all. I scooped up handfuls of cold, clear water, splashing my head, dousing my hat and staring at the sun until my eyes went black. I sat down, facing the creek's flow, and then lowered myself down to my elbows so that the water ran up my chest and into my nose. I could dig here, I thought, but I would find nothing. Perhaps, I thought, it was time to move on.

Then, as surely as the person who picks the same numbers for each week's lottery, I hauled myself out of the water, fetched my shovel and began to dig.

I had to. What if this was the week that my numbers came up?

13

Nearly Normal Norman and the Corrosive Power of Gold

Tom's luck did not improve. While he was going cold turkey he developed an abscess – possibly more than one – underneath his remaining teeth.

'I can't believe it,' he said, rolling around in his bunk, unable to find peace. 'I decide to give up one of the strongest pain-killers known to man and I get toothache.'

Craig and I had been worrying about Tom for days. We would visit to see how he was, take him coffee and cold drinks and offer food, but he wasn't hungry. I had developed a deep respect for Tom; very few people have the courage or strength to withdraw from opiates without assistance – but to withdraw from them whilst in extreme physical pain? That was something else.

We rounded up other, perhaps inappropriate, painkillers from everyone in the camp – diclofenac, codeine, Vicodin, hydrocodone, Advil, co-codamol – but he turned them down. I tried to find oil of cloves but drew a blank. Tom went to the town's clinic and was given a course of antibiotics, which he had to complete before the offending tooth, or teeth, could be pulled.

'I need to get out of here,' he said one day. 'I want to go back to Oregon. Terry told me about a place on the Rogue River where they're pulling out good gold. You can still dredge in Oregon – no legal problems. You want to come?'

Good gold.

Back in the original Gold Rush, it was commonplace for miners to abandon their diggings in one place at the first whiff of a rumour of richer pickings in another. Often, the rumours would be false; and even when they were true, those who weren't fast enough on their feet would find everything staked and claimed by the time they arrived. In 1850, for example, more than ten thousand headed north to the Trinity River, a tributary of the Klamath not far from where I was prospecting now, on hearing that it was laced with placer gold. It was probably no richer than the rivers the miners had left in the south, but the promise of big finds was irresistible.

Tom wanted to go north to Gold Hill, a small town in Jackson County, Oregon, but I had told myself that if I moved I would head south, down to the rivers that ran through the Mother Lode. This was the area to which my friends William Swain, Sarah Royce and John Borthwick would head. It was gold eroded and washed from the Mother Lode that the first 49ers picked up from the ground. As a mother feeds her babies, it was still feeding the rivers with placer gold way down there. I was three hundred miles away from its northernmost tip and that was eating me up. But Terry knew what he was talking about. Good gold. I began to waver.

'You get better and we'll think about it,' I said to Tom.

Evenings at Big Foot were calm and quiet, cool but not cold before their midnight tipping point. Craig, Marcy and I would sit enveloped in the shrill sound of cicadas and talk into the night, and I felt this was healing time; I would forget about gold while my frayed muscles knitted into knots that would hurt in the morning. Neither Craig nor Marcy were interested in prospecting. I stopped describing my days to them; I became afraid of Marcy's disapproval.

Marcy was a free spirit and she always would be. Originally from Baltimore, she married and divorced the same man twice, the divorces coming in 1969 and 1971. 'There were lots of other people after that, but no marriage,' she told me. 'I lived in Maryland for a long time on a farm and when I hit thirty these hippies started showing up. I found them so attractive and wonderful and thought, "That's how I want to be."

'I had always felt like that when I was married. Even if I listened to the Rolling Stones my husband would say I was too young – but they were my age! We belonged to the country club and I would spend my time with the black bartender rather than the boring rich white people. I never felt that I fitted in until I met these hippies.

'Eventually I took off with this other girl and we hitch-hiked all the way to Arkansas, to this beautiful town called Eureka Springs. It was delightful. It had lots of hills and contours and old Victorian houses. As soon as I got there I began jumping up and down. I felt I had finally found my home.

'Almost straight away I fell in love with a guy called Nearly Normal Norman. This is a man who was so whacked out he used to stick a finger into an electric socket when he wanted to clear his brain. I stayed with him for six or seven years before breaking up, then I bought a forty-acre place out of town on a mountainside. Shortly after that I met Ron, a Cherokee, who was the sperm donor for my daughter Willow. He turned out to be an asshole.

'We got caught growing pot when Willow was just three months old. There was so much stuff – eight quarts of bud all dried and in jars. I couldn't think where to hide it. I was there with my daughter while outside Ron was talking with the police. They took Ron away and three days later they came back for me. I had to hand Willow over to another woman for three hours while I was questioned; it was the worst three hours of my life.

'They charged us with growing eighty-five plants but seventy-five were just seedlings. It was a small town and small-town cops are mean. Eventually, the judge let us off with a five-hundred-dollar fine and one year of unsupervised probation. It was 1979.'

Ron took off with another woman and Marcy carried on living up the mountain, bringing up Willow. She said it was peaceful and idyllic, but it wasn't to last.

'One night a guy just walked into the house,' she said. 'He had heard about the marijuana and thought I might still have some. Well, Willow was there and I got protective like a mother bear, screaming, "Get outta my house!" and I moved back into town.'

After that, Marcy described a peripatetic life of packing up and driving to places – the Florida Keys, back to Arkansas, the town of Truckee in California, then Michigan, Oregon and Mississippi – always with the long-suffering Willow in tow. Willow was in her mid-thirties when I met Marcy, but when, in her twenties, she had met a football coach and told her mother she wanted to settle down in Arizona, Marcy had been heartbroken.

Marcy's journey did not end there – she relocated again and again, seemingly retreating deeper into isolation – and when I met her I wasn't sure that Happy Camp was its conclusion. I had to remind myself that she was seventy-four. Willow wanted her mother to move closer to Arizona but she seemed to be resisting that. She didn't like Arizona – and feelings of guilt were clouding her judgement.

'As I get older I feel terrible about what I did to Willow,' she told me one night under the soft red glow of my lamp. 'I dragged that poor girl from pillar to post. Every time she settled in a new school or made new friends, I would have an idea about some other place and we'd move on. It was so selfish.'

Almost every night, before anyone else came to call, Marcy would sit next to me and ask about some aspect of British or European life. She was inquisitive and funny, and if time had been on her side I'm sure she would have been moving faster, longer, but you can't keep moving forever.

While I was deciding whether to move on, Dave Mack invited me for a beer at the New 49er headquarters. The building was closed and quiet, lit by artificial light – all the windows were blacked out and shuttered – and I was surprised to see that he had a windowless bedroom there. He lived here during the summer season but had a home in the Philippines when he wasn't prospecting in other parts of the world.

I might have been wrong, but he seemed lonely to me in the way that leaders of men, or cults, often are. His first wife had died of breast cancer long ago and a second marriage had ended in what he said was a friendly divorce several years earlier. 'I just wasn't home enough,' he told me. He had a thirty-one-year-old daughter named Christine but he was living on his own. I had the feeling, not for the first time, that rooms seemed smaller when he was in them.

In common with the early miners, McCracken described disputes with the Karuk and accommodations that he had made in order not to offend them, not least steering clear of Ishi Pishi Falls, the centre of their spiritual universe. But there were constant disputes over dredging and use of the river. I found that each side had gripes about the other (and many Native Americans shared his passion for gold).

We talked about his adventures for two hours before I asked him head-on why he had devoted his life to gold.

'Did you ever read *The Hobbit*?' he asked. I told him I had. 'It's all about the ring, *the ring*. Tolkien had it right about all the traits and the corrosive power of that ring and the deep desire to possess it. It has a simple meaning – even the best of men could not hold that ring for long without being seduced by it. Raw gold is just like that.

'When you find it, particularly underwater when you are sucking gravel and the gold follows behind and you have to dig it out, the moment of discovery is something on a level with religious experience, every single time. It does something to you. It draws out strong feelings of exhilaration from deep down inside of you. You could take a PhD in chemistry but I don't think you'd ever figure out the chemistry of what happens to you the moment you realize you have uncovered gold.

'It isn't the same with silver or with a wad of hundred-dollar bills or a casino slot machine payout or a win at poker. Because gold is pure wealth – beautiful, shining – exposure to it does something more to you than make you want to possess it. In that moment when you uncover it you have uncovered something beyond money. It is something that is akin to pure freedom, not only in a material sense, because gold transcends the US dollar or any other paper currency. You can convert it to any currency at market price. It is portable, and that makes you free. People want to buy that gold. I can take it anywhere and I will have people fighting over who is going to buy it. In a material sense, I transcend whole political systems by possessing that gold. I am actually bringing it out of the earth where no man has seen it before. It is like being a conduit that brings pure wealth out of the ground, and that is something special, almost spiritual. Not in the same sense as going down on my knees to pray, but in a different way that's like a rush of sheer excitement and meaning.

'Once you have taken gold from the earth, you change. Perhaps it's like someone who smokes crack cocaine for the first time. One inhalation and your life changes forever. You won't feel that good unless you do it again.'

Privately, I suspected that McCracken was right but I never voiced my thoughts. We sat in silence for a while, me shocked by his candour – him, I suspect, by his unguarded honesty. Money is vulgar, gold is beautiful, I thought. Or was it?

'I'm sixty years old,' he went on. 'I am the most aggressive underwater gold miner in the world and I don't need money. I can retire any time I want, but what would I do? I still search for gold. Why? For that moment of discovery. The feeling you get when you find raw gold – I love it. It's a feeling of joy and exhilaration and surprise.'

In common with most of the miners I encountered, Dave Mack was happy to concede that he had gold fever. Given the definition of it as *a mania for seeking gold*, he was unlikely to deny the fact. The machinations for finding gold might be prosaic and exhausting but the moment of discovery was almost indescribably exciting and that was the hit, the rush. But transcendent? Spiritual? I wasn't so sure, at least not yet.

What I really wanted to know was whether it was possible to contract gold fever without succumbing to its darker symptoms and I was coming to the conclusion that, certainly over the long term, this might be as likely as falling just a little bit pregnant.

'I work with very few people because eventually gold ruins friendships, and there are only so many of those that you can afford to lose in one lifetime,' McCracken said. 'I've seen best friends, brothers even, fighting over gold discoveries or over who should get a share of what. I've seen people killed over it. I've had a tiny number of partners over the years because I only ever found a very few prospectors who understood the destructive power of gold the way I do, who know what to look out for, how to read the signs.

'The only way to keep from losing yourself is to be of a solid mindset, determined that everyone involved will be given their full share of whatever is recovered. Gold never corrodes but you won't find anything on earth, anything at all, that is so corrosive.'

I never saw Dave Mack again. The next day, I told Tom that I would go with him to the Rogue River, and the idea that I would be travelling on a whim, a rumour, on the promise of 'good gold' like some gullible loser in 1849 was not lost on me. It had been my intention to remain in California but a visit to Oregon would give me the chance to try looking for gold with Tom's dredge; it would probably be my only opportunity, as dredging had been banned everywhere in California except Siskiyou County, the very place that I was leaving.

Tom had begun to join us again in the light of my red lamp, though not for long periods. He was feeling better with each passing day but it was clear he was still weak and in pain. Marcy was delighted to see him – Millie had a soft spot for Tom – but she didn't approve of our proposed expedition.

'You're damn fools,' she said. 'What do you hope to find there?'

'Gold,' said Tom. 'Terry told me about a place—'

'Sure he did. And you're all going to get rich.'

Marcy included Craig in this but we were not expecting him to join us. He was more interested in white-water canoeing and had previously shown no desire to go prospecting. But then he suddenly said, 'Can I come too?'

Craig was originally from Douglas, Arizona, and had been taken to Los Angeles by his mother when he was about five years old. His father had abandoned his mother when she was six months pregnant; Craig had never met him.

'I'm vaguely interested in meeting him,' he told me one night. 'But he knew I was out there. He was an adult, and that says something about him. I couldn't imagine doing that if I had a son, so that means he's a lesser person than me. Do I want to associate with him? No.

'There was time as a teenager when I thought a lot about him, but I decided it was the wrong thing to do. The man I regard as my father married my mother when I was six and then four years later they had my brother and I guess I was pretty much ignored after that.'

Craig loved his brother and felt jealous and resentful of him at the same time, and that kind of dichotomy is difficult for a kid to handle. There were five half-sisters, too.

'I became bad trouble,' he said. 'I had a hugely troubled young life. They took me to see shrinks and counsellors and I saw what they wrote about me – that I was "troubled" or "mildly psychotic" or "bipolar". When I was seven I pushed a six-year-old kid into some bushes – it was a stupid thing to do but his mom called the police and I was arrested and charged with assault and battery. I had a record at the age of *seven*.

'This kinda carried on and at thirteen I was put into a psychiatric hospital for six months. I was pretty out of control. Looking back, I think I was mainly just a highly strung, intelligent, hyperactive kid, and they just didn't know what to do with me.

'My stepfather had gone on to divorce my mom and she was working three to eleven every night and so I saw her for around fifteen minutes a day. The only people I had to keep an eye on me were five half-sisters and my half-brother, but when it came to them I felt like an outsider and I acted out on that, kicking back at it. I did everything to my half-brother short of punching him in the face. I was pretty bad to him and I regret that now. We're not friends. The only ones I'm friends with are my second oldest sister, Deanna, and fourth oldest, Lori.

'Lori was really Mom and she made sure everyone was fed and our homework was done, though in my case homework was pretty much non-existent – at school the teachers gave me assignments Monday and Tuesday but I would get bored and wouldn't go back until Friday, when there was a quiz. My work would be A-standard but half the time I would end up without enough hours in class to get a grade.'

I found all this difficult to accept because I felt Craig deserved better. He had a sharp intellect and a ready wit. That summer, on a whim, he had taught himself the laws and concepts of hydraulics. If anything broke down in the camp, he could fix it. He was constantly scrounging books and would often press me on the oddest concepts at two in the morning at my makeshift table.

'Why don't they do 3D printing in space?' he asked me one night. 'They could print huge stuff. There are limits to how big you can print things on earth, but they would be gone up there . . .' He clicked his fingers '. . . just like that.'

We had both been drinking. I stared at him. He was smiling and I knew there was a flaw in his argument. So did he. He was just waiting for me to find it.

'Because you need gravity,' I said. 'Otherwise the stuff you're printing with would just float off everywhere.'

We stared up at the stars. I found Orion and wondered if I would ever drift south again.

I slapped my leg. 'You could create centrifugal force in a spinning space ship! You could print in there. That could be your gravity.'

'Exactly,' said Craig, triumphantly. 'And you could print the spinning space ship.'

'Where?'

'In space.'

'But . . .'

He had attended a party at a friend's house when he was nineteen. Later, he heard that someone had tried to climb into bed with the friend's mother after the party had finished. The next time he saw the perpetrator, Craig beat him up. He then assumed he would be wanted for assault and so he took off, seeking out farming and service jobs in Wyoming, Colorado, Idaho, always accepting cash, never revealing his identity.

It was five years before he found out he wasn't wanted at all. 'I know it sounds crazy,' he told me, 'but I think I didn't want to know the truth. I was having too much of a good time.'

When he was twenty-eight, Craig married a woman named Christine, who was four years older than him. She suffered from brittle diabetes, a type that affects fewer than one per cent of diabetics and results in debilitating and wild oscillations of blood glucose levels. It doesn't have to be fatal, but where it isn't managed properly, it can be. It can result in depression and psychological problems, and in some sufferers this can result in a dark sense of inevitability.

'One of the first things we did when we got married was to sit down and watch a film about a couple in our situation,' Craig said. 'And she said, "This is exactly what is going to happen. You will come home one day and I'll be dead on the floor. It could be six months, it could be five years."'

He was thirty-six when she died.

'I was in the Los Angeles County Jail at the time,' he said. 'I can't remember what for – something assault-related. My next-door neighbour called the prison chaplain and he came and told me the news in my cell. They would have given me a compassionate pass, but because I was in jail when she died,

Christine's family cut me off. I didn't know where the funeral was, or when. Instead, I just stayed in my cell and cried for three days. I still don't know whether she was buried or cremated or where she is.'

I could tell that Craig was embarrassed by this. I hadn't known him when his wife had died, but I could tell he was a better man now.

He'd remarried in 1997 but that had recently ended. Twelve months ago, he had been in a motor bike accident that cost him his job, and he'd had a lot of time to think about life while recovering from a broken back. He had been here all summer in an RV that he bought with his savings, and when I asked why here, he told me that he and his second wife had visited Happy Camp on their honeymoon.

He didn't know what he was going to do or where he was going to go next, but I had no doubt that he would make a fortune when the right opportunity came along.

'Don't you think it's a bit weird coming back to the place where you spent your honeymoon?' I asked.

'No,' he said. 'Not at all.'

14

The Unluckiest Man
in California

I rose early and felt at once excited by the journey that lay ahead and sad for the one that was coming to an end. I thought of Sarah Royce and doubted that she had felt the same way; the first part of her Gold Rush adventure had been much more arduous than mine. I screwed up my tent, emptying disturbing quantities of rocks and gravel from it, threw my clothes into my rucksack, folded and stowed my garden chair and airbed with lackaday carelessness, and threw everything else into the back of the car, prioritizing my mining equipment and beer cooler.

Tom was already up, having tied his dredge to the rear of Priscilla with the kind of skill that one might reasonably expect from a professional diver. He was still in pain but his spirits were much improved. The antibiotics had not worked and now he was on a second course, meaning that his teeth still had not been removed.

'I'll feel better when we get on the road,' he said. The more

I got to know Tom the more undeserved his bad luck seemed. I was willing it to change.

With a spring in my step I went around the camp saying goodbye to everyone, most reluctantly to Rita and Gary, Marcy, Terry and JoAnne. I had said a provisional farewell to the latter couple the previous evening when I bought one of JoAnne's pendants. In turn, she passed me a small vial containing my share of the gold mined during our day out together; I could see now why they weren't wildly impressed by the gold-covered rock I had found. The pinpoints of colour on it might have shone deceptively, making my heart beat faster, but once washed off they were little more than tiny specks of brilliance that didn't amount to much. Still, I was grateful and added them to my stash.

It was a cool seventy-seven degrees and the day was bright and clear, the blue of the sky accentuated by a few disintegrating wisps of cloud. It would grow much warmer but we would not care as we headed north, over the mountains and into Oregon.

I had read enough to know that I should have been questioning the wisdom of this trip, one of the sad characteristics of many of the losers in the Gold Rush being that they had a tendency to flit, often fruitlessly, between supposedly rich sites. But it would give me an opportunity to see a corner of southwestern Oregon, which had its own place in the original Gold Rush; significant deposits were found there in 1850. Placer miners had rushed up from California and worked on Josephine Creek; also on the Illinois, Applegate and Rogue rivers, Port Orford and Ellensberg on the coast, where gold was found in black sand near the estuary of the Rogue River. After that, Ellensberg came to be known as Gold Beach.

We were headed for Gold Hill, which was upriver in Jackson County. Gold was discovered there in 1849 by miners heading south to the California Gold Rush, but it wasn't thought to be worth bothering with. Later, in 1857, a pocket on the top of a small mountain was found to contain upwards of a ton of gold. It was, quite literally, a gold hill.

I checked that my gold was safe in my shoulder bag, and then I checked again. Craig loaded a bag and his inflatable canoe into Priscilla and we set off, heading out of Big Foot in a convoy of two, miners waving at the side of the road. Rita gave us a teary wave with a dusting cloth as if we were embarking on a journey on board the Orient Express. Whatever the coming days held for me, I knew there would be no one to fill the shoes of the irrepressible Rita. I waved back.

We turned left on to Grayback Road and plunged into the cool shade of a million overhanging branches. Craig was sitting up with Tom and I drove behind them as we began climbing into the mountains of the Siskiyou National Forest, Priscilla only occasionally achieving her top speed of 30mph, me trying hard not to shunt into the back of her.

Our plan was to head over the mountains and to continue north-east on Highway 199 to Grants Pass, then due east on Highway 99 to Gold Hill, which was situated on the north bank of the Rogue River. It should have taken about four hours but we'd reckoned without Tom's luck and Priscilla's age. We had been on the road for less than fifteen minutes when Priscilla had her first blowout, the nearside rear tyre reduced to shreds in an instant. We were preparing to find the spare wheel when a passing forest ranger pulled over with a huge trolley jack at the ready. We were on our way again in less than twenty minutes. Perhaps Tom's luck was changing at last.

Once we crossed into Oregon the road descended gently and the terrain softened. Down we went to green pastures and broad flats occupied by red Dutch barns and small farmhouses squared off with white picket fences. The roadside was punctuated by colourful painted signs offering blueberries and apricots, roses and lavender, tomatoes and blackberries. After weeks of glorious mountain vistas this was a new, different, beauty and it was easy on the eye. We stopped for Tom to check Priscilla's tyres, and Craig switched to travel in my vehicle for a while. Then he rejoined Tom at the next check. This was typical of Craig; he didn't want either of us to be alone for too long.

It was a full one hour and twenty minutes before Priscilla had her next blowout. We were on the outskirts of Grants Pass, a small but busy town named after General Ulysses S. Grant, when I realized I had left my friends at some traffic lights. I tried to double back but lost them and it was a full seventy minutes before I located them at a tyre repair shop, scratching their heads and replacing the wheel.

'That's my seventh blowout in ten weeks,' said Tom.

'How is that even possible?' I asked.

'It's the wheels. They're 16.5 inch rims and nobody makes tyres for them any more. I thought I was lucky when I got these.'

'We should find the guy who sold them to you,' said Craig, darkly.

'They were given to me,' said Tom.

'Huh?' I said, uselessly. 'So we can't hunt him down?'

'Nope.'

After this, Tom had no more spare tyres. Craig and I were in favour of remaining at Grants Pass and searching junkyards for remnants from the 1970s but Tom wanted to push on to Gold Hill. This meant he was about to drive on to a major highway in an antediluvian vehicle on tyres that were probably more than forty years old, tyres that had been remoulded more times than Joan Rivers and which were carrying more weight than they could now reasonably bear.

'What if you have another flat?' asked Craig. 'What then?'

'I'll be fine. 'We'll just have to be careful, go slower.'

While Tom was stowing away what was left of the last blown

wheel, Craig leaned towards me. 'Do you think he makes his own luck?' he asked.

'Does anyone?' I replied.

We limped into Gold Hill after a precarious journey, Priscilla weaving like a fat man on Quaaludes, me behind her flashing my hazard lights and trying not to stall.

The town had been established after the discovery of gold on Big Bar along the Rogue River. Previous to that, it had been known among Native Americans as a mysterious and forbidden place from which their horses – in fact, all animals – would shy away. No one knew why. It had a similar population to Happy Camp but, to my mind, was infinitely less attractive. One main road ran through it, on which were housed a saloon, a few fast-food joints, a large grocery store and not much else. I felt as repulsed by it as a horse might have been.

We pulled over for a beer and a slice of pizza at the sort of soulless restaurant that makes teenagers want to leave town as soon as they can. The one thing that Gold Hill did have in abundance was RV parks for summer visitors who wanted to try their hand at prospecting or white-water rafting, but Tom had been to the town before and said they were expensive and usually full. Tom and Craig were broke and I wasn't far behind, so we decided to find a legally dubious free option at the side of the road, even if it meant being arrested, having our vehicles impounded and, in the long run, paying much more than an RV park would have cost.

Once it got dark I found myself wondering why I had ever considered leaving Happy Camp for this most unhappy settlement. On the edge of town and close to the Rogue River, Tom found a dusty lay-by and we drove into it, pushing as far as we could into trees so as not to be seen from the road. We were not supposed to stay there for the night but we were past caring. Should cops with flashlights wake us up, we had enough dumb faces and foreign accents to get us out of trouble.

While Tom prepared Priscilla for him and Craig, I lowered the rear seats of my four-by-four, threw everything under the vehicle and pumped up my mattress; I would sleep in my own car: it was cleaner, newer, and I figured that fewer creatures were living in it.

The air became cold and moist. We had a quiet nightcap and cheered up at the thought of hauling the dredge down to the water in the morning. I would be weighing myself down with a hundred pounds of lead and jumping into the fast-flowing waters of the Rogue River with a respirator in my mouth and a huge vacuum hose over my shoulder. How I had changed in a few short weeks.

Tom went to bed and Craig passed me another beer. He would help us to carry the dredge to the river but after that he would take off in his canoe.

'You know what?' he said, rhetorically, taking a swig on his drink. 'There's something wrong with you guys.'

I had considerable difficulty extricating myself from my sleeping quarters the next morning, and my travelling companions seemed reluctant to help me. My inflatable mattress had lost some air and I had sunk, like a gold nugget, to its centre as it had noiselessly wrapped itself around me.

'He looks like a nacho,' said Craig through the open window.

'Very funny,' I muffled. 'Get me of here.'

Tom was standing in Priscilla's doorway, laughing. It was the most relaxed I had seen him in some time and I felt pleased for him. He was cooking bacon and eggs and I suddenly realized I was ravenous. I managed to open a door and crawl from my rubber trap, falling into the thick grass at Craig's feet.

'Morning, sport,' he smiled. He was already drinking and he smelled of marijuana. He was on his own road trip.

I made my way through the trees to take a pee and slowly came to, realizing I was urinating down a mineshaft whose wooden cover had rotted and collapsed in on itself. Tom and Craig came to look when they heard my shouts.

'That'll be some old gold diggings,' said Tom.

'Shall I go down there?' I asked, wondering again why I hadn't brought a rope.

'Sure, if you want to get killed,' said Craig. 'You don't know what's down there.'

'But whoever dug this might've left some gold.'

'Like I said,' muttered Craig. 'There's something wrong with you.'

Our side of the bank, the northern side, was covered in rocks and willows on whose branches was debris suggesting that the waterline was usually much higher than it was now. There wasn't much prettiness about this side. Across the river were houses half-hidden by oak trees on a high bluff, some with ladders leading down to jetties. One had a small elevator to a boat down below. After the unspoiled beauty of the Klamath, it all seemed rather shabby.

The day was warm and the sun shone sporadically through a ceiling of flat, dun clouds. After breakfast, we began assembling Tom's suction dredge, with Tom and Craig arguing over the best way to do this, and we dragged it down to the bank of the Rogue River. I knew what was coming next.

'You're going in,' said Tom.

He proceeded to dress me in a thick rubber wetsuit, flippers and mask. He handed me the respirator and stood me in water up to my chest.

'The pump will give you air on demand,' he said. 'Just try to breathe normally. The lead weights will get you down to the riverbed. The current will keep trying to pick you up and throw you down the river, but the weights and hose should keep you on the bottom. Grab rocks if it gets too fast. Keep on sucking up sediment, and hopefully we'll find some gold.'

Craig was reclining on a garden chair, happily taking photographs, drinking beer and grinning. 'You guys . . .' he kept saying. He stopped only to take off his shirt, apply some sun lotion and take another beer from the cooler.

I thought of Dave Mack. *'Three minutes . . .'* he had said.

The water where we were was much more sedate than that in which Dave Mack and Rich Krimm had laboured on the Klamath, but it had a current that was far too powerful for me to counter in the event that I might need to – I'm a poor swimmer.

I had put to the back of my mind the fact that two of the three lead belts that Tom had attached to me were home-made. He had brought one of his own weighing forty-five pounds from Australia, but on arrival in America he found that the price of lead had sky-rocketed. To get round this, Tom had acquired some lengths of rubber fire hose, into which he had stuffed hundreds of tiny lead weights of the kind used to balance wheels. He had found these at a junkyard and now the hoses that encased them rested over my left and right shoulders, crossing my chest. I looked like a bandit without a gun.

'Remember,' said Tom. 'Just breathe normally.'

I made a circle with the thumb and forefinger of my right hand, sending Craig into paroxysms of laughter, and plunged into the icy water.

At first it was cloudy and brown, but as the weights dragged me to the riverbed my view became clearer and my confidence grew. I dragged myself and the vacuum hose against the current into deeper water and felt a vibration as Tom switched on the suction pump from the bank. I was limited in how far I could travel by the length of the hose attached to the floating dredge, but I estimated I was thirty feet from it when I hit the middle of the river at a depth of around ten feet.

Here there were small boulders and contours that I figured could have trapped gold as it was driven by the power of past floods. I edged into a hole about three feet deep and began sucking up sediment through my giant vacuum, being careful all the while to avoid trapping the crayfish that were scuttling away from me. The current was relentlessly powerful and I strained to fight against it. If I could make this hole deeper, then I could work inside it and shelter from the fast-flowing water at the same time. Down I went, four inches, six inches every few minutes.

I was surrounded by feeding fish drawn by the disturbance. The water pummelled my shoulders like a brutal masseur but the sense of struggle was exhilarating; being weighed down underwater, unable to surface without crawling to the bank, had seemed terrifying at first, but now it felt oddly safe. Without this weight, I would become a piece of wood tossed around by an unyielding current. My breathing steadied as I heaved and tugged the suction hose. I imagined flakes of gold – nuggets even – trapped in my vortex, flying up to the dredge.

I lasted ten minutes, then twenty, twenty-five, and felt calm, in control. Shafts of sunlight rippled through the darkness and I entertained myself by following them. One set something ablaze in my peripheral vision and I twisted quickly to see what it was, knowing instinctively as I turned, and as first one weight belt, then the other, fell from my shoulders, that it could be only one thing: fool's gold.

In an instant, my centre of gravity had moved from my sternum to my hips. My upper body, still exposed to the rush of the river but suddenly lighter, was thrust backwards while Tom's home-made weights slipped farther down to my thighs, calves, ankles. I dropped the suction hose and tried to catch them, my body now in the shape of a scoop that the current hit like a jack hammer, wrenching me out of the hole and sending me into a spin that peeled back my face mask and filled it with brown water.

With the loss of two weight belts, I was now too light to resist the current. But with one buckled around my waist, I was too heavy to swim to the surface. I was struck by the depressing obviousness of this whilst spinning around and around, unable to see beyond a welter of bubbles that seemed to be floating and then sinking.

I tried to right myself but it was hopeless. Blindly, I wrestled with the buckle around my waist but my rubber-encased fingers were as deft as a troll's. As I spun on the axis of my weight belt, riding the current downstream, my shrinking world became dark then bright, dark then bright. I wasn't sure whether the speed of my rotation slowed or my perception of it did, but suddenly it didn't seem to matter – or mattered less. I felt terribly relaxed. While I still had the respirator in my mouth, I could breathe, I thought, drowsily; what was the point of fighting?

It couldn't have been that I was unconscious – it was more like a trance – but the jarring of the respirator as I ran out of line seemed to wake me up, the force of the jolt almost pulling out my teeth. Now that the air line from the dredge was taught, I had something on which to hold, and I hauled myself, underwater, back to the shallows.

I emerged exhausted, gasping for breath, and wrenched off my mask. Tom had a grave look on his face. He could see that something had gone wrong but he just patted my back and said, 'Well done.'

Three minutes, I kept thinking.

I flopped out of the water and on to the sandy bank. I lay on my back for a few moments, breathing fresh air, and stared at the sky, trying to put a value on what had just happened.

'I think that's enough for today,' said Tom.

Farther up the bank, Craig offered me a beer. I shook my head.

'Hey, I got some great shots,' he said, uncrossing his legs and handing me the camera. I wondered whether he had filmed my dilemma or taken stills. Either way he couldn't have recorded much – a flailing arm, perhaps, or a flipper spinning out of control.

When I had caught my breath, I reviewed his handiwork, scrolling through stills and frozen frames one by one. Only then did I discover that while I was beneath the surface a group of women in bikinis, rafting over the white water, had sped above me, oblivious to the small drama playing out below. Craig had pictures of them all.

That night we cooked lamb on Priscilla's two-ring stove. I drank wine while Craig and Tom enjoyed a few beers, Tom's alcohol-free. We avoided talking about gold. Craig had brought a computer and so we found ourselves by the Rogue River in Oregon sitting in a lay-by and watching a comedy. Periodically, someone from one of the houses on the other side of the river would shout at us to shut up.

In the morning Tom was in pain again and so I drove into town to look for a dentist who would treat him. He had the offending tooth removed while Craig and I waited in a diner. He appeared, groggy and haggard, after an hour or so.

'How do you feel?' I asked.

'So-so,' he said through numb lips. 'I need to lie down.

Worst of it is this: see this lump in my neck?'

'Yes.' I had noticed it before and had assumed from its size
– like a marble – that Tom must already have had it checked
out.

'The dentist says I need to see a doctor about it immediately. I'll try to find one tomorrow.'

No one used the word 'cancer' but no one needed to.

'Shall I find you a doctor today?' I asked.

'No. I'm too tired.'

I was desperate to head south to the goldfields I had heard
so much about. It was clear there was to be no more dredging
today or tomorrow. Maybe, given Tom's condition, there
would be none at all for some time to come.

I drove us back to Priscilla and put Tom to bed. Craig and
I sat outside and debated what to do next. He had work to do
for Rita at Big Foot and I had plans of my own, plans that
involved looking for gold. I made a virtue of offering to drive
Craig back to Happy Camp and, after several hours, when
Tom said he was feeling better, we said our goodbyes and left
without seeing any colour.

As I sped south towards the Mother Lode I imagined Tom
lying again alone and in distress, worrying about his diagnosis
and wondering why his friends had left him, but I put the
thought to the back of my mind. I had things to do, gold to
find, and my heart was as hard as the diamonds I had lied
about at Somes Bar.

Tom went home to Australia without finding any gold. He
was diagnosed with cancer and wrote to tell me he was confi-
dent he would beat it.

Sure he would, I told myself. But only if his luck changed.

PART II

'Next to the Civil War in the 19th century, no other event had a greater impact, more long-lasting reverberations, than the Gold Rush. It transformed obviously California, but more importantly, it transformed America.'

J.S. Holliday, historian

In California

When the Argonauts finally made it to California they encountered a pace of life, a vitality that no one – perhaps anywhere else in the world – had experienced before on such an international scale. The summer, autumn and winter of 1849 saw the arrival of some ninety thousand gold-seekers, and they were all in a hurry. Wherever gold was found, swarms of miners would follow. Entire towns – canvas tents at first, then wooden cabins, then brick buildings – would rise and fall, sometimes in a matter of days, sometimes in weeks or months, as gold was found and exhausted. At first, there were no sheriffs, no courts and no jails, but rates of serious crime were surprisingly low – perhaps because in the absence of such authorities the noose was the quickest way to dispense justice.

With wives left at home and without the moral constraints imposed by community and church, men found themselves free to do whatever they wanted, whenever they wanted. With little else to pass the time outside their diggings, drinking and gambling became the main pastimes and they were pursued with vigour. It was not unusual for a miner who had been in the goldfields for months to come into town and lose all his wealth overnight. If you were abroad at sunrise in San Francisco or Sacramento you might catch sight of him heading shamefaced and crestfallen back to the mountains.

Fairly soon, the smartest Argonauts realized that the fastest way to get rich was to mine the miners. Prices – usually weighed out in gold dust – were astronomical. If you take it that hitherto a dollar was about the average daily wage, imagine the joy of the woman who made eighteen thousand during a spell in the goldfields by baking and selling pies; or the foresighted man who arrived in San Francisco in July 1849 with fifteen hundred old newspapers, which he sold for a dollar each to miners, hungry for news from home.

The writer Bayard Taylor arrived by ship in the summer of 1849 and feared that no one would believe him when he described the San Franciscan economy in dispatches he had been commissioned to send to the *New York Tribune*. When the average wage for a labourer in New York was one or two dollars a day, it might be hard to accept that individual hotel rooms were rented to professional gamblers for upwards of ten thousand dollars a month. The proprietor of one of San Francisco's first hotels, the Parker House, leased his entire second floor as a gambling den for sixty thousand a month.

Taylor wrote: 'A friend of mine, who wished to find a place for a law office, was shown a cellar in the earth, about twelve feet square and six deep, which he could have at two hundred and fifty dollars a month [the equivalent of 7,500 today].

'[One] citizen of San Francisco died insolvent to the amount of forty-one thousand dollars the previous autumn. His administrators were delayed in settling his affairs and his real estate advanced so rapidly in value meantime that after his debts were paid, his heirs had a yearly income of forty thousand dollars. These facts were indubitably attested; everyone believed them, yet hearing them talked of daily, as matters of course, one at first could not help feeling as if he had been eating "of the insane root".'

Those who arrived early in the Gold Rush found any kind of covered accommodation to be extremely scarce, and prices reflected this. Taylor ended up paying forty-five dollars a week for room and board.

'Every newcomer in San Francisco is overtaken with a sense of complete bewilderment,' he wrote. 'The mind, however it may be prepared for an astonishing condition of affairs, cannot immediately push aside its old instincts of value and ideas of business, letting all past experiences go for naught and casting all its faculties . . . Never have I had so much difficulty in establishing, satisfactorily to my own senses, the reality of what I saw and heard . . .

'A curious result of the extraordinary abundance of gold and the facility with which fortunes were acquired, struck me at the first glance. All business was transacted on so extensive a scale that the ordinary habits of solicitation and compliance on the one hand and stubborn cheapening on the other, seemed to be entirely forgotten. You enter a shop to buy something; the owner eyes you with perfect indifference, waiting for you to state your want; if you object to the price, you are at liberty to leave, for you need not expect to get it cheaper.'

Taylor was describing a perfect sellers' market. During the early days of the Gold Rush, which were characterized by shortages, a miner might be expected to pay up to a hundred dollars for a pair of boots (equivalent to three thousand dollars today); forty dollars for a pound of coffee; fifty dollars for a pick axe; three dollars for a dozen eggs (though one miner reported seeing them on sale for a dollar each); two dollars for a loaf of bread; and sixteen dollars for a tin of sardines. One Argonaut described an eating establishment in one camp charging a dollar for a slice of bread, and two dollars if it was buttered (equivalent to sixty dollars today). Edward Gould Buffum, author of *Six Months in the Gold Mines* (1850), describes taking a breakfast of bread, cheese, butter, sardines and two bottles of beer with a friend and receiving a bill for forty-three dollars (equivalent to about thirteen hundred dollars today).

Gold was to be found in great quantities – certainly in the early days of the rush – yet many miners were finding only enough to cover their costs. What use was it if you could literally find gold in the mud of San Francisco's streets (Taylor says that five dollars' worth a day could be harvested this way, dropped by 49ers returning from the mountains) if it was eaten up by a room, a drink and a meal?

When Taylor arrived in 1849, San Francisco was buzzing with activity, yet still very primitive, largely built of canvas and wood. By the time John Borthwick disembarked there in the spring of 1851, there had been many changes for the better. Food and accommodation were not as scarce and you could dine at a variety of restaurants as good as anything you might find in New York or Paris. If you had money or gold you could go to any one of hundreds of saloons and gambling houses, or 'hells' as they were called, to lose it; you could even go to the theatre.

'Good hotels were not wanting, but they were ridiculously extravagant places,' wrote Borthwick. 'And though flimsy concerns, built of wood and not presenting very ostentatious exteriors, they were fitted up with all the lavish display which characterises the fashionable hotels of New York. In fact, all places of public resort were furnished and decorated in a style of most barbaric splendour, being filled with the costliest French furniture and a profusion of immense mirrors, gorgeous gilding, magnificent chandeliers, and gold and china ornaments, conveying an idea of luxurious refinement which contrasted strangely with the appearance and occupations of the people by whom they were frequented.'

San Francisco was bawdy, vibrant and filthy, Borthwick

recalled, and was 'famous for three things – rats, fleas and empty bottles.'

It was customary for miners to rent a bed in a 'room' separated from others by nothing more than a canvas wall, and to dine out. Driven by boredom, many would then seek company in the gambling dens and simply get drunk. Drunkenness was endemic. The journalist Frank Soule wrote that San Francisco's drinking culture was 'the worst feature of the city':

> The quantity of ardent spirits daily consumed is almost frightful. It is peddled out in every gambling-room, on the wharves, at almost every corner and, in some streets, in almost every house.
>
> Many of the taverns are of the lowest possible description – filthy dens of vice and crime, disease and wretchedness. Drunken men and women with bloated bodies and soiled garments, crowd them at night, making the hours hideous with their bacchanalian revels.

By the mid-1850s San Francisco had 537 registered saloons, one library and an army of prostitutes – some accounts estimate as many as eight thousand.

The prices of goods would fluctuate wildly with cycles of scarcity and plenty. It took only the arrival of a ship loaded with flour to see the price of bread plummet, and it was in this economic landscape that entrepreneurs conducted business at a sprint. Remember those hundreds of ships abandoned in San Francisco harbour (mentioned in Part One)? Many were commandeered by restless merchants and turned into warehouses and hotels. Others were simply sunk, filled in and used as the foundations of new wharves and landing stages as the city expanded into the bay. The speed of growth was unprecedented. Risk-takers, convinced that only success was possible in such a climate, borrowed money at interest rates of up to fifteen per cent per month.

Borthwick described the maelstrom, breathlessly: 'In the course of a month, or a year, in San Francisco, there was more hard work done, more speculative schemes were conceived and executed, more money was made and lost, there was more buying and selling, more sudden changes of fortune, more eating and drinking, more smoking, swearing, gambling and tobacco-chewing, more crime and profligacy and, at the same time, more solid advancement made by the people as a body, in wealth, prosperity and the refinements of civilisation, than could be shown in an equal space of time by any community of the same size on the face of the earth.'

Argonauts who came overland stopped short of San Francisco, arriving first at the goldfields beyond the Sierra Nevada Mountains to the east. Those who did not remain permanently in California would see their first glimpse of the city as they made their way home by sea; few chose the overland route back, once more ships had arrived with cheaper passages and regular timetables.

Sarah Royce's first experience of a mining camp was at Weaverville, west of Redding, where she, Josiah and Mary stayed for two months. She was relieved to be across the mountains but understandably concerned for the security of her family.

'The sense of safety that came from having arrived where there was no danger of attacks from Indians, or of perishing of want or of cold on the desert, or in the mountains, was at first so restful that I was willing, for a while, to throw off anxiety,' she wrote, 'and, like a child fixing a play-house, I sang as I arranged our few comforts in our tent.'

Yet these concerns, it seems, were soon to be replaced by others: 'Still, there was a lurking feeling of want of security from having only a cloth wall between us and out of doors. I had heard the sad story (which, while it shocked, reassured us) of the summary punishment inflicted in a neighbouring town upon three thieves who had been tried by a committee of citizens and, upon conviction, all hung. The circumstances had given to the place the name of Hang-Town. We were assured that, since then, no case of stealing had occurred in the northern mines; and I had seen, with my own eyes, buckskin purses half full of gold-dust, lying on a rock near the roadside, while the owners were working some distance off. So I was not afraid of robbery; but it seemed as if some impertinent person might so easily intrude or hang about, in a troublesome manner.'

Josiah tried his hand at panning for a while, but he enjoyed little success. He had some experience of trading and this resulted in a small group of men offering to go into business with him in the establishment of a store. Sarah was not unhappy at the prospect – she perhaps saw such an enterprise as socially more acceptable than mining – and was fully supportive of her husband as he set off for Sacramento with his investors' money to buy supplies from – who else? – Sam Brannan. (I visited Brannan's store, a grand three-storey brick building in historic Old Sacramento, and felt unjustifiably disappointed to find it was now a bicycle shop.)

While Josiah was away, Sarah tried to employ men to build her something more substantial than a tent, but no one could be persuaded to leave their diggings. When Josiah returned and the store was opened, she saw up close the full spectrum of types who had made it to California as they came in to buy provisions.

'The majority of them ... were men of ordinary intelligence, evidently accustomed to life in an orderly community where morality and religion bore sway,' she wrote. 'They very generally showed a consciousness of being somewhat the worse for a long, rough journey, in which they had lived semi-barbarous lives, and for their continued separation from the amenities and refinements of home. Even in their intercourse with each other they often alluded to this feeling, and in the presence of a woman, then so unusual, most of them showed it in a very marked manner.

'But, mingled with these better sorts of men who formed the majority, were others of a different class. Roughly-reared fron-tier-men almost as ignorant of civilised life as savages. Reckless bravados, carrying their characters in their faces and demeanor, even when under the restraints imposed by policy. All this and more were represented in the crowd who used to come for their meat, and other provisions in the early morning hours.

'There were even some Indians who were washing out gold in the neighboring ravines, and who used to come with the others to buy provisions. It was a motley assembly and they kept two or three of us very busy; for payments were made almost exclusively in gold-dust and it took longer to weigh that, than it would have done to receive coin and give change. But coin was very rare in the mines at that time, so we had our little gold scales [and] weights, and I soon became quite expert in handling them. While thus busy, in near communication with all these characters, no rude word, or impertinent behaviour was ever offered me.'

Fairly soon, Sarah saw – and remember, these are still rela-tively early days in the Gold Rush – signs of disappointment and despair. Many Argonauts had been through traumatic experiences and witnessed death and extreme hardship. They missed home. They missed female company and the simple joy of seeing children at play, because there were hardly any there. During 1849, a total of 697 ships had landed forty-one thou-sand people in San Francisco, just eight hundred of whom were women. The proportion of women arriving overland was even smaller.

'A number of miners passed every morning and afternoon, to and from their work, but none of them stared obtrusively,' wrote Sarah. 'One, I observed, looked at Mary with interest a time or two, but did not stop, till one day I happened to be walking with her near the door, when he paused, bowed courteously and said, "Excuse me madam, may I speak to the little girl? We see so few ladies and children in California, and she is about the size of a little sister I left at home."'

Sarah felt kindly disposed towards the man. '"Certainly," I said, leading her [Mary] towards him. His gentle tones and pleasant words easily induced her to shake hands, and talk with him. He proved to be a young physician who had not long commenced practice at home when the news of [the] gold discovery in California induced him to seek El Dorado, hoping thus to secure, more speedily, means of support for his widowed mother and the younger members of the family.'

The physician's company was not untypical – it included a lawyer, a geologist and other 'cultured gentlemen'. Many others, Sarah noted, were mechanics, farmers and merchants, and while their knowledge was doubtless invaluable in the success and formation of the new state, one can only imagine how their absence affected the smooth running of small towns and villages across the rest of the United States.

'Sounds of discontent and sadness were often heard,' Sarah noted. 'Discontent; for most of them had come to California with the hope of becoming easily and rapidly rich; and so, when they had to toil for days before finding gold, and, when they found it, had to work hard in order to wash out their "ounce a day"; and then discovered that the necessaries of life were so scarce it took much of their proceeds to pay their way – they murmured; and some of them cursed the country, calling it a "God forsaken land," while a larger number bitterly condemned their own folly in having left comfortable homes and moderate business chances for so many hardships and uncertainties.'

The sounds of sadness were 'deeper and more distressing', caused by sickness and death. Many had seen friends and loved ones die along the way and they were deeply disturbed by their losses. Others died from a simple lack of understanding of basic hygiene and nutrition (Sarah described a group of young men who fell desperately ill after travelling all the way from Missouri without once bathing or eating fruit or vegetables. She said that at least one of them had died.) It was probably Sarah and Josiah's own superior knowledge of these basics that preserved them when they each contracted a bout of cholera the month after their arrival.

Argonauts arriving late in 1849 or early 1850 found that most of the gold-bearing land had already been claimed. During the whole of 1848 it is estimated that a maximum of six thousand people went looking for gold in the Sierra Nevada. By the end of 1849 there were more than forty thousand – with a larger number on the way – so arrivals would often find that there was nowhere for them to dig. When this happened, they had four choices: they could either work as a labourer for a day rate on another miner's claim, a prospect that stuck in the craw; they could move on in search of fresh claims, the option most often chosen; they could buy a claim, which was risky, expensive and often fraudulent – it was not unusual for vendors to 'seed' claims with a sprinkling of gold for unsuspecting purchasers to find during their first inspection; or they could go home, and thousands did.

Those who returned empty-handed, often in debt to friends who had remained in the goldfields, were said to have 'seen the elephant', a wonderful Americanism that means taking something positive from a negative experience. The most often cited provenance for the expression involves the hick farmer who travels to town to see the circus, only to have his cart upended and all his produce ruined when his horses are spooked by a parade of the circus animals. Instead of being annoyed, the farmer is elated. 'I don't give a damn,' he says. 'I have seen the elephant.'

By January 1850, William Swain and his friends had set up camp on the South Fork of the Feather River 'twenty-five miles from Long's Trading Post and sixteen miles above Bidwell's Trading Post' in the northern reaches of the Sierra Nevada.

Writing to his brother, George, William sounds an optimistic note, perhaps because he and his friends had heard that the South Fork had not been much worked or perhaps because he wanted to shelter George and his wife, Sabrina, from the truth – that men were dying and gold was hard to find.

One miner wrote that many men had rushed into the high goldfields too soon and that during the winter months 'hundreds have been stricken down by disease; many died, while others have been unfitted for work for the rest of the season . . . When a man gets sick in the mines, even if he has a physician and medicine, the food he gets is not of the kind required and prices of medical attendance and of necessities are so high that a month's sickness sweeps off a big pile.'

The few doctors in the goldfields were unlikely even to take a man's temperature without the promise in advance of an ounce of gold.

Writing on 6 January, William is more positive: 'After prospecting for two days, we located a spot favourable for damming and draining the river. We made our claim and then built a house as soon as possible to shelter our heads from the soaking rains. So here we are, snug as schoolmarms, working at our race and dam. Whenever the rain will permit, a fall of the river will enable us to get into the bed of the river and know what is there.

'If there is no gold, we shall be off to another place, for there is an abundance of gold here, and if we are blessed with health, we are determined to have a share of it.'

16

To the High Sierra

After leaving Tom, I sped south to the Mother Lode, where most of the original 49ers, including William Swain, John Borthwick and, for a time, Sarah and Josiah Royce, had staked their claims.

I canvassed the opinions of several experienced miners on where might be the best place to start and settled on the northern stretches of the Yuba River. Someone had told me about a guy who had found a nugget *this big* in the Yuba only a week before my arrival. Someone else had said that long stretches of the best-known gold river, the American, could attract too many prospectors during the summer months – and the last thing I needed was people. In common with the 49ers, I had realized that people meant one thing: competition for whatever gold was present. After seeing only a handful at Gold Hill in Oregon, I decided that, on the whole, I preferred solitude.

I raced back through Yreka and retraced my steps down Interstate 5 through Shasta City, Dunsmuir and Redding, aiming the car, with characteristically flawed logic, at Yuba

City, some three hundred miles away. Here, I thought, without any research whatsoever, would be a town with a large hole at its centre from which the Yuba River would flow impossibly, spreading out into shallow creeks that would accommodate gentle gold-mining activities.

In fact, Yuba City is primarily associated with the business of farming and packing fruit in the Sacramento Valley. The land on which it stands is at the confluence of the Yuba and Feather rivers about fifty miles north of Sacramento and was originally owned by our old friend John Sutter. He sold it to some enterprising businessmen at the start of the Gold Rush. By 1852 about two hundred people were living there, enjoying the amenity of a post office, one hotel and a grocery store. Today, it has a population of sixty-five thousand.

I made a brief visit to the town's very pretty museum and learned an awful lot about peaches and multiculturalism and a little less about gold mining. An exhibit there told me that twenty-five thousand Chinese had flocked to the region during the Gold Rush and they had their own name for California: Gam Saan, meaning 'Gold Mountain'.

I jumped back into my car and found a road that meandered east through rolling countryside and fruit orchards. Near an old mining town called Smartsville, named after a Gold Rush hotelier named Jim Smart, I saw a sign at the side of the road that read 'Action Mining Supply'. Where there were mining supplies, I thought, there would be mining operations. I turned the car down a dirt track and followed it until it wound on to what looked like a small farm with a wooden industrial building, outside of which stood two men apparently concluding a transaction.

They stopped to look at me.

'You got mining supplies?' I asked through the car window.

Wordlessly, the men looked at one another, then at the sign above the door that read 'Action Mining Supply', then, blankly, at me. They finished their business and I had the impression that the customer left sooner than he had originally intended.

'Howdy. How can I help you?' asked the proprietor.

I was looking for a suction bottle, rather like a turkey baster, which I had been advised would be useful for sucking sediment out of underwater crevices. 'Have you got one of those?' I asked.

'Sure, come inside.'

The proprietor's name was Hoss Bundy and what he didn't know about gold prospecting wasn't really worth knowing. He was aged sixty-five and, in common with almost every man I met on my gold travels, he wore blue jeans, checked shirt and a baseball cap. He had a white beard and brown-framed spectacles and on his counter top was a copy of the *Pick & Shovel Gazette*, the newspaper of the Gold Prospectors Association of America (GPAA). After a few moments of small talk, Hoss took me into the back of the store and showed me his collection of nuggets and pickers, flakes and dust, and demonstrated the best methods for separating gold from quartz, should I find any so inconveniently packaged by nature.

Having just left a hundred miles of rivers and creeks that I had been given carte blanche to prospect by Dave Mack, my biggest concern was over mining rights and claims. Where, I asked Hoss, could I legally look for gold? He thought about this for a moment, taking into account my obvious status as a novice. Mining clubs such as the New 49ers and the GPAA had large tracts available to members, but I wasn't a member.

'Well, there's plenty of places you can go on public land, national parks and such, but they tend to be cleaned out already,' he said. 'The best places to find gold are usually on private claims and you're taking your life in your hands if you work someone else's claim illegally.'

I gulped. 'And these are all clearly marked?'

'Some are, and some aren't. It's up to you to do the research.

I heard of people being shot for working claims without permission.'

I could see this was going to be a problem. Not much had changed since 1849. Without clear signs of a claim, it would be irresistibly tempting to park up at the side of a quiet creek in the middle of nowhere and begin to dig. Cue the sound, after several hours, of a pick-up truck coasting to a halt at the top of the bank, a rifle cocked, a dog growling as it slips its leash . . .

Of course, Hoss had his own claims but he told me how he worked others too, without having his head blown off. It was very simple.

'Beer,' he said.

Dreamily, I was about to say 'Yes please,' when he went on. 'You see some guys working a claim and you walk down to their diggin's with a six pack of beer. You've already opened a cold one and you're sucking on it and going, "Ah, that hit the spot," and then you offer the guys a beer and ask if you can give them ten or twenty dollars to prospect on their claim for the day. If they've been working hard, they usually take a beer and invite you to work upstream or downstream. Some will just let you on for a beer, some might take the money too, and others might ask for fifty per cent of whatever you find, just to cover themselves in case you hit a good pay streak. Don't take too much beer, though, or you mightn't do any work. Give it a try. They might not even ask for money.'

I thought this was excellent practical advice and I thanked Hoss, but not before asking one more question.

'Where should I go?'

He thought about this for a while. 'Try Downieville,' he said. 'Good a place as any.'

I was delighted with Hoss's suggestion because John Borthwick had visited Downieville in 1851 and found it a pleasing and strangely cosmopolitan destination – even though it was among the highest and most inaccessible towns on the

Mother Lode. I had intended following Borthwick there at some point and this seemed the perfect opportunity.

After thanking Hoss and buying another bucket and a steel tool for eking out the contents of crevices, I continued east, driving through Grass Valley and Nevada City and on to Highway 49 towards Downieville.

I defy anyone to tell me that California State Route 49, the 'Mother Lode Highway', is neither the state's prettiest road nor its most practical, linking, as it does, eleven counties and the majority of the Mother Lode mining towns from Oakhurst in the south, through names with which I had by now become intimately acquainted and dangerously obsessed – Mariposa, Sonora, Columbia, Jamestown, Angels Camp, Mokelumne Hill, Sutter Creek, Placerville, Coloma, Georgetown, Auburn, Colfax – and on northwards beyond Downieville and Sierra City.

These were all places of great discovery, heroic enterprise and breathtaking endeavour. To travel the entire 295-mile length of Route 49 – as I was to do in the coming weeks – is to experience some of the most outstanding scenery on the continent, from regal mountainsides, crystal lakes and fearsome rivers to peaceful hamlets, bountiful orchards and windswept acres of farmland by the thousand.

To traverse Route 49 and to imagine the inaccessibility of the landscape before the road existed is also to recognize and marvel at the resourcefulness of the hardy 49ers who gave it its name.

I approached Downieville from the south and west, the road hugging the North Yuba River as it sliced into the Sierra, hewing out deep canyons with the most persistent of white water. It was surrounded by peaks covered in pine and fir and, closer to the river, oak and maple trees. At three thousand feet, the town is one of the highest and most isolated of the Mother Lode region – and if you didn't know that, then you might have guessed it as you breathed in. Exposing one's lungs to clean, rarefied air is similar to drinking mountain spring water; it turns something unremarkable into something transcendent. There was no horizon to be seen anywhere from Downieville; mountains in all directions saw to that. In its heyday, the town had been alive with energy and hope.

Borthwick had prospected and sketched in Downieville and had found himself enamoured of its inhabitants and the many tongues they spoke. It was rough and wild, while enjoying a certain sophistication that belied its isolation. He had heard many stories about Downieville gold finds, but when he first approached it – small and surrounded by mountains as it was – he had found it disappointing. It was not until he arrived that its charm took hold of him.

'I very soon discovered that there was a great deal compressed into a small compass,' he wrote. 'There was only one street in the town, which was three or four hundred yards long; indeed, the mountain at whose base it stood was so steep that there was not room for more than one street between it and the river.

'This was the depot, however, for the supplies of a very large mining population. All the miners within eight or ten miles depended on Downieville for their provisions, and the street was consequently always a scene of bustle and activity, being crowded with trains of pack mules and their Mexican drivers.'

As the town was situated at a fork of the Yuba River, miners called it 'The Forks'.

'The diggings at Downieville were very extensive; for many miles above it on each fork there were numbers of miners working in the bed and the banks of the river,' recalled Borthwick. 'The mountains are very precipitous and the only communication was by a narrow trail which had been trodden into the hillside and crossed from one side of the river to the other, as either happened to be more practicable; sometimes following the rocky bed of the river itself and occasionally rising over high, steep bluffs.'

In 1849 conditions in Downieville had been harsh and during winter food was hard to find; many miners died there. By the time Borthwick arrived, however, mule trains from Marysville, and everything they carried, had made the town a much more pleasant place to live and work. He enjoyed 'luxurious' lodgings run by a Frenchwoman and was well fed and even entertained in the town's small theatre.

'The most prominent places in the town were, of course, the gambling saloons, fitted up in the usual style of showy extravagance,' he wrote. 'There were several very good hotels and two or three French restaurants.'

A company of American glee singers performed every evening to cheer up the miners and one night, Borthwick explained, 'Their selection of songs was of a decidedly national character, and a lady, one of their party, had won the hearts of all the miners by singing very sweetly a number of old familiar ballads, which touched the feelings of the expatriated gold-hunters.' So much did it touch them, in fact, that a collection for her raised five hundred dollars (worth about fifteen thousand today).

The town itself was named after another Scot, William Downie, a sailor who was given the title of 'Major' by his compatriots even though he seems not to have had any military experience. It was more a term of endearment and an acknowledgement, perhaps, of his skills as a leader. Born in Glasgow in 1820, young Downie was something of an adventurer, travelling to Australia, Canada and the East Indies before pitching up on the east coast of the US. Here, in 1849, he first heard about the discovery of gold in California. With none of the indecision that had slowed down other would-be Argonauts, he boarded a ship immediately in Boston and was in San Francisco by 27 June, well ahead of the overlanders. Sarah

Royce had begun her journey from Council Bluffs just nine-teen days earlier and would not reach California until 19 October.

After a brief spell prospecting at Marysville, across the Feather River from Yuba City – where I had been that morning – Downie had headed for the hills and, at the confluence of the North Yuba River and its north fork, today called the Downie River, he had found gold, and lots of it.

Writing about his adventures in *Hunting for Gold* in 1893, Downie described arriving there:

The scene that burst upon us was one of marvellous beauty, and after these many years it still lies before me like a lovely panorama, in my recollection of the moment when I first saw it. The silence of the woods was broken only by the rushing of the meeting currents below and the soughing of the breeze through the foliage. The sun was in the western sky, causing a variation of light and shadow to fall upon the landscape, which was exceed-ingly pleasing. The hillsides were covered with oaks, bending their crooked branches in phantastic forms, while here and there a mighty pine towered above them, and tall willows waved their slender branches, as it were, nodding us a welcome . . .

Down on the very brink of the river grew a beautiful grove of fir trees, and as we approached, a frightened deer ran from the thicket and made for the woods. Near a little spring, which bubbled up and made the surround-ings look fresh and verdant, stood a few pieces of bark on end – the only sign that human foot had ever trod this region, and further indicating that here at some previous time the Indians had camped.

Add to this the waters leaping over rock and boulders, and the clear azure sky stretching like a canopy over the whole landscape, and you have the picture, as far as I can describe it, that I first beheld, when I approached the Forks.

Downie and his party of nine arrived in October 1849 and, wisely, quickly built a cabin in which to spend the winter. They began prospecting to pass the time and soon found that they were able to fill a cup with gold each day, a feat that earned the location the name of Tincup Diggins. By the summer of 1850 word of their success had spread and more than five thousand miners were working claims in and around The Forks. Two miles north of the town, some lucky Argonauts found a nugget that weighed twenty-five pounds. In one claim measuring just ten feet by six, another group found twelve thousand dollars' worth of gold (the equivalent of about 350,000 dollars today) in just eleven days.

Riches poured out of the ground at other spots named Poker Flat, Sailor Ravine, Whiskey Diggins and Blue Banks. Famously, one housewife found five hundred dollars' worth of gold when she examined the sweepings from her kitchen floor.

I found a deserted campsite ten miles out of town. My spot was surrounded by fir and maple trees and behind my tent was a path that wound down to the North Yuba some twenty yards distant. Here, beyond a fringe of willows, the river flowed slowly to a depth of about one foot, perhaps two at its middle, and I was to find it perfect for bathing in the absence, anywhere, of showers or washing facilities. I would sit on a rock and douse myself slowly and deliberately at the end of each day's prospecting, watching the sun slip behind magnificent ponderosas and wild vines on the opposite bank. Bears were often seen here and I hoped that I might spy one across the water at some lucky dusk, but I never did.

On the first night I lit a fire and read by its light. I was too excited by the next day's possibilities to feel hungry, but I ate some cold beans anyway, temporarily disposing of the tin in a bag hung from a tree some distance from my tent. If I were to have visitors, I preferred them not to disturb my slumbers.

This, I thought, was exactly what I had wished for when I began my journey. I had learned how to look for gold and how to find it. Now that I had the isolation in which to put my new skills to the test I would not let myself down, because if I wasn't doing this for me, who was I doing it for?

Clueless in
Downieville

So far, I had cared more about learning how to prospect for gold than actually finding it. The minuscule amount I had in my vial was excusable because I had spent my time in education. When I woke shortly after dawn, I vowed to change this. I was in an area known to bear good gold and, feeling as insistent as William Swain, I was going to pocket my fair share of it. I would work hard and for long hours. I was alone; there would be no distractions.

I dressed quickly, made myself a peanut butter sandwich and climbed into the car. It was a beautiful morning, cold but bright and filled with opportunity. As I left camp I saw a notice-board and stopped for a few moments to read a flyer that had been pinned to it. 'PLAGUE CAUTION', it read: 'Chipmunks, ground squirrels or other wild rodents in this area may be infected with plague.'

I later heard a radio report explaining that the plague referred to was of the bubonic variety. The sign went on, 'You can get plague by: The bite of an infected flea; touching or holding an infected rodent; being around a pet cat infected with the plague. Early symptoms of plague include fever, chills, headache, muscle aches and swollen and tender lymph nodes. If you develop any of these symptoms within seven days of your visit to this area, let your doctor know that you have been in an area where rodents may have plague.'

What next? I wondered. Idly, I thought how annoying it might be to survive an attack by a bear – possibly one that had been spooked by a rattlesnake hiding in a grove of poison oak – only to be bitten by a plague-infected flea shaken from the animal's fur.

I followed the North Yuba heading into Downieville, leaping slightly from the driver's seat each time I saw a perfect spot near the river where a discerning miner might find gold. And each time I did, my eyes would pan from the river far down below to the trees lining the roadside and I would see notices pinned to them declaring that this stretch of river or that had already been claimed. It seemed to me that all of it belonged to somebody. I slowed to a halt underneath one claim that bore the name of an individual. Others had no names at all. Some had the names of mining clubs of which I was not a member. This, I thought, was a depressing and unfair state of affairs: to come all this way; to be so near, yet so far. For a moment, I thought of the thousands of miners to whom this had happened in 1850 – many arrived in California, half dead, only to find that the 49ers had already claimed everything worth working.

The worst of it was that I could not see anyone actually digging or panning on the claims, which meant that moseying on down to any of them like Hoss Bundy with a six pack of beer was out of the question. I drove into town and saw a tiny tourism office, at the door of which was an official who told me I was in luck; there was a guy there – just three feet away

– called George, who knew everything there was to know about gold mining in and around Downieville. Fortune was smiling on me – perhaps he was even *the guy*. The two men whispered for a while, and then George stepped forward.

'What can I do for you, young man?' he asked, myopically.

I explained my desperate situation and desire not to be shot, run over or mauled by hunting dogs for claim-jumping. 'Where,' I implored, 'could I safely look for gold?'

'Hell, anywhere,' he replied.

'Anywhere?'

'Sure. Most o' them claims are bogus anyway.'

'Bogus?'

'Yeah. Some o' them haven't been worked for years. Paperwork's out of date.'

'Which ones?'

'You'd have to find out from the Bureau of Land Management.'

The nearest land management office was ninety miles away. I found myself staring silently at George.

'Pan wherever you feel like,' he said. 'You get caught, just act dumb.'

This was easy for George to say. He was a local man and the population of Downieville was just 282. He probably knew whose claims he could safely work, especially if his friendly neighbours were the owners. They wouldn't be so friendly to me.

'Thank you for the advice,' I said. I might have been mistaken, but I could have sworn George's eyes lit up a smidgeon. As soon as I left he would be on the phone to his fellow miners. 'Load up the dogs,' he would say. 'We got a live one.'

The best advice I was able to canvass, from a forest ranger (the area fell inside the Tahoe National Forest), was that it was fine to prospect along the river as it flowed through the 'city limits', the 'city' being about three hundred yards long, and at a publicly owned area called Union Flat, a twenty-minute drive farther east. Surely, I thought, anything in or near town would have been cleaned out by the locals as soon as this year's floods had abated. I would head eastwards.

Union Flat came with a campground that was similar to my own and, similarly, it was empty. It was a Monday and later I came to find that, even in summer, some isolated campsites like this were completely depopulated from Monday through to Friday morning and utterly overrun with booming, often drunken weekenders from Friday to Sunday afternoons. They would come with fifty-foot RVs and satellite television sets playing loudly through open windows above the noise of their screaming children. The air would vibrate to the sound of snarling dogs, the chink of empty beer bottles and the shrill sizzle of twenty-four-ounce dinosaur steaks overcooking on gas-bottle barbecues the size of small countries.

Several times when silence on Thursday was replaced with commotion on Friday, I packed up everything, drove farther up into the mountains and slept in the back of my car, returning on Monday morning to camp in peace.

I gathered up my equipment and climbed over a bank above the river. A duck, disturbed by my footsteps, flapped from the undergrowth and panicked its way to the other side, disappearing into thick oak and pine trees. The Yuba was shallow and slow at this time of year and its passage was almost musical as it climbed and fell over rocks and stricken tree trunks. The sun had risen and the temperature was already above ninety degrees but I took comfort at the sight of clouds pitching slowly above, providing cover as I waded downstream towards a bend in the river. I was acutely aware that dozens of other miners would have seen this bend too, just this year, but I was limited in my choices by claims and a lack of local knowledge. And, anyway, I was luckier than them, or smarter, or harder-working. Perhaps my eyesight was better. I had some edge, I just knew I did, which would make my endeavours bear more fruit than theirs.

On the inside of the bend I found a group of rocks just above the waterline that appeared unmolested and decided I would dig there. I dunked my hat in the river, put it on and began hewing down to bedrock.

The day grew intensely hot but the river cooled the air around me and I found myself whistling. When, I wondered, was the last time I had whistled? Mid-morning, when I had two buckets of concentrates ready for panning, I walked upstream to a stretch where the river was deeper and I dived in. I floated on my back for a time, fully clothed, watching birds criss-crossing the sky, heading higher into the sierra, and thought of Schrödinger's cat, the animal in a box that quantum physicists argue might be dead and alive at the same time. While my concentrates remained in their buckets, unpanned, I might be rich or poor. Already, experience had told me to enjoy the uncertainty while it lasted.

I prepared one more bucket of concentrates and stopped for lunch – another peanut butter sandwich and an apple – whilst sitting on a rock with my feet dangling in the cool, clear water of the Yuba.

During 1850 this river would have been alive with activity. The Forks had been a difficult place to access and even harder to supply but it exploded into life in just a few short months. Many of the five thousand who arrived by June did well; others did not. And come the winter, it is said that many who had not made provision for cold and hunger – whether through stupidity or lack of funds – died in their flimsy tents. Even by the standards of other goldfields, prices here were exorbitant because of its isolation. Until regular mule train routes to the outside world were established, you could expect to pay three dollars for a pound of potatoes and up to 150 dollars for a pair of boots (the equivalent of about 4,500 dollars today).

I spent the afternoon panning my concentrates and averaged a flake in each pan. It was paltry reward for my efforts but at least it was colour and it gave me confidence. I had come alone, chosen a spot alone and found gold, alone. More experienced prospectors would have laughed at my lack of success – and if they called me clueless I would find it difficult to argue with them.

But I wasn't as clueless as I used to be.

Downieville is one of a long and creditable list of towns along Route 49 that might appear in a movie about the 1849 Gold Rush with hardly any cosmetic changes at all. If you were a film director you could decide on a whim to drive in and begin shooting immediately. Just move the cars out of the way, spread a little sawdust and, 'Action!', you would be transported to the 1850s.

As John Borthwick noted, Main Street is not very long but history crowds around it, making the town seem bigger and more substantial than it really is, not least because it has no fewer than three bridges with just two hundred yards dividing them all. Many gold-era buildings remain, among them the museum, which used to be a Chinese store and gambling house, and the big brick Craycroft Building, originally built in 1852 as a saloon with a seventy-foot-long bar, which gives you some idea both of the numbers populating the town and their priorities. Over the years the Craycroft has served as a courtroom, restaurant, newspaper office and jail.

There are lots of pastel-coloured wooden buildings with ornate balconies and tin roofs. My favourite from that era was the pink home of the *Mountain Messenger*, California's oldest newspaper. It first appeared in a mining camp called Gibsonville and was originally published as the splendid-sounding *Gibsonville Trumpet*. Several times I ate dinner at the Grubstake Saloon, one of a handful of watering holes in town, and serving as a placemat on my table would be a copy of a *Messenger* front page with a splash that read:

THE GREAT YUBA PASS CHILLI COOK-OFF: Neither Rain Nor Snow, Nor Sequestration, Nor Terrorist Attack ... Sierra County's Premier Social Event on Schedule.

That's the spirit, I thought.

Sadly for Downieville, the town is best remembered for the most shameful day in its history, the day that a mob hanged a pregnant woman – the first and last woman to be hanged in California. It was July 1851 and the prospectors of the fledgling town had been enjoying their Independence Day celebrations. A group of them, led by one Frederick 'Frank' Alexander Augustus Cannon, had been enjoying them rather too much. After drinking past closing time in the town's saloons, Cannon is said to have tried, with some friends, to break into the home of a Mexican woman, Josefa Segovia, with the intention of raping her.

What happened next is not entirely clear, but the next day Josefa stabbed and killed Cannon in her home, apparently in self-defence, when he returned, some say to rape her, others say to apologize. A kangaroo court was hastily assembled, comprising mainly Cannon's friends, and Josefa was found guilty of murder and sentenced to death, in spite of evidence from a Dr Cyrus Aiken that she was pregnant. The fact that she came to be named 'Juanita' in subsequent accounts of the case is, perhaps, indicative of the atmosphere in which she was convicted; she was Mexican so few troubled even to name her properly, simply calling her something generically Latino. Since the loss of California to the United States, Mexican citizens had been treated with precisely the kind of contempt that resulted in this young pregnant woman being hanged from the town's Jersey Bridge, with hardly anyone to speak out for her.

William Ballou, a passing miner who arrived in Downieville just hours after the hanging, wrote: 'I had just got to the edge of the town when looking down the grade I saw something on the bridge looking like a piece of [cloth]. When we came down we found it was a Spanish woman that had killed a man. The mob took her out and hung her. It was the first woman I ever saw hung, and it was the most degrading sight I ever saw.'

Borthwick arrived a few weeks after the hanging and, from his account, it appears he was fed a sanitized version of events: 'A Mexican woman one forenoon had, without provocation, stabbed a miner to the heart, killing him on the spot,' he wrote. 'The news of the murder spread rapidly up and down the river, and a vast concourse of miners immediately began to collect in the town. The woman, an hour or two after she committed the murder, was formally tried by a jury of twelve, found guilty and condemned to be hung that afternoon. The case was so clear that it admitted of no doubt, several men having been witnesses of the whole occurrence; and the woman was hung accordingly, on the bridge in front of the town in the presence of many thousand people.'

News of the hanging spread far and wide, heaping shame on the people of Downieville. Why had they not defended Josefa and her innocent unborn child? It was the question on everyone's lips, not just in America but, eventually, across the Atlantic. News of the hanging reached Britain in just under two months – quite a feat – when *The Times* in London carried a report from the San Francisco *Picayune* under a headline that read, 'Lynch Law on a Woman':

A party of drunken men, reeling home after a debauch, forced the door of a private house in which a woman resided, and insulted the inmates. The house was not a brothel, nor the woman a prostitute. Of this outrage no notice was taken, but, on the party returning for the vilest of purposes and insulting the woman with the grossest epithets, her southern blood boiled over with indignation and, in a moment of passion, she snatched a knife and inflicted upon him a mortal wound. For this she was taken to the crossroads and hanged!

Now we venture to say that had this woman been an American instead of a Mexican . . . had she been of the Anglo-Saxon race, instead of being hung for the deed, she would have been lauded for it . . . It was not her guilt which condemned this unfortunate woman, but her caste and Mexican blood.

Having little regard for the reputations of those it saw as the guilty parties in the case, the *Picayune* thundered:

The judge – John Rose of Rose's Ranch – let his name be remembered – stated to the witnesses that they must tell the truth 'just as if they had been sworn'; to which they sentenced and lied; a physician who testified on her behalf was beaten, a lawyer who endeavoured to get justice done for her was threatened with hanging and ordered to leave town, and amid the shouts of bloodthirsty monsters, whose rage could hardly be restrained, even for one short hour, the upright, intelligent, and impartial jury, brought in a verdict of 'Murder', and condemned her to be hanged in two hours – which sentence was executed!

One circumstance was wanting to make the murder of this woman unparalleled in atrocity. She was in a condition that would have made her life sacred, even in the most barbarous ages, and under the bloodiest code that ever cursed the world. An unborn infant perished with its murdered mother, before it saw the light!

Some people say that Josefa haunts the streets of Downieville. Others say she doesn't haunt it alone and that her baby, in the shape of a glowing orb, accompanies her. On several evenings, I found myself the last person abroad on Main Street and I confess to being spooked on more than one occasion, but it wasn't Josefa and her child who I imagined lighting my way.

No, in the thickening night, a wind whistling through the steel frame that supported the town's bell, I envisaged stepping from yards and alleyways Judge Rose, Frank Cannon and his acolytes, looking for someone else to hang. If anyone were doomed to walk these streets 'til the end of time, I thought, quickening my pace, surely it would be them.

'NO, NO, NO, NO, NO!'

I was hitting myself over the head with my gold pan on a stretch of the Yuba on the outskirts of Downieville. If ever you feel the need to rattle your cranium with a piece of plastic, I would have no reservation in recommending this location for the job. There are huge rocks that look like the folds of a giant woman's skirts, tinkling rivulets of cold clear water and crows that settle themselves along branches on the far bank, hoping that your labours might bear fruit, drawing blood, perhaps, and leaving you in an unconscious and vulnerable state in the shallows, the better for pecking at.

I had not been enjoying much luck. Day after day I would rise early, pee in the bushes, dunk my head in the river, look for bear tracks around my tent, prepare a peanut butter sandwich, jump in my car and set off for whatever diggings I had chosen on a map by the light of my fire the night before. This choosing was always a lengthy process as my fear of accidental claim-jumping had not abated. I was usually breaking soil somewhere by 7am and aimed to have enough concentrates by lunchtime to keep me panning until mid-afternoon, by which time the heat usually spelled the end of my day. In short, I was not shirking; I was working hard but I wasn't finding much gold.

I harked back to what Rich Krimm had told me way back in Happy Camp: 'You can work as hard as you like, but if the gold isn't there, no amount of work will make you lucky.'

The most frustrating aspect of this was that I regularly found one – just one – flake of gold when I had carefully washed everything else out of my pan. A more seasoned gold-seeker would have called this prospecting, not mining. This was the process whereby you dug and panned in a few areas to decide whether the gold you could get out of the ground was worth your while. They would have washed a few pans, seen a flake each time and emitted a judicious 'Nah'. But I was new to this and figured that one flake was better than nothing. In reality, I was wasting my time but I was still entranced by each new sighting of colour.

Today, I had emptied a crevice about twelve yards in length, three feet deep. The result was nine or ten almost weightless – and therefore almost worthless – pieces of gold.

When I'd stopped hitting myself over the head, I dipped my hat in the river and put it on. The crows flew away.

I blamed William Downie for my lack of success. Not the tens of thousands of miners who had picked through this area in the years since he'd arrived at The Forks. Just him. Just Downie. I had decided I did not like 'Major' William Downie. This is because in his *Hunting for Gold* he presents an account of the hanging of 'Juanita', first damning her with faint praise – 'She was proud, and self-possessed, and her bearing was graceful, almost majestic' – before going on carefully to craft a description of the incident that avoids heaping further shame on the town that bears his name.

Not only does he say that harmless old Frank Cannon had fallen drunkenly through the door of Josefa's house in the dead of night, accidentally and in a manner befitting a theatrical farce, neither waking nor abusing her, but also that Cannon was stabbed by Josefa the next morning when he returned, puppy-like, to apologize. Downie makes no mention of her being pregnant. Instead, he goes on to suggest that the poor woman actually *hanged herself*.

'Calm and dignified she mounted the scaffold,' he writes. 'There was a death-like silence in the crowd, everybody wondering what she was about to do. Logan [the hangman] seemed involuntarily to surrender the rope he was supposed to place around her neck, and with her own soft hands she placed the noose in position.

'"*Adios Senors!*" she said with a graceful wave of her hand, and ere the astonished spectators could realize what had happened, she had leaped from the scaffold into eternity.'

Pah! I thought.

And if that wasn't bad enough, Downie and his cohorts had scooped up all the gold that was supposed to be waiting for me. 'On Zumwalt Flat we went to work with a rocker and the first day washed out about twelve ounces,' he writes. Then, later, 'I dug a hole near a small bar . . . [at] Tincup Diggings and found it would pay about one dollar to the pan.' That was the equivalent of a regular day's pay in each pan washed.

Infuriatingly, Downie adds: 'We had difficulty weighing our gold. Some claimed to be making from thirteen to fourteen ounces a day, crevicing in the banks . . . [We] found gold all along the banks, sometimes several hundred dollars within the short space of a few hours, very seldom using even a shovel. Our principal mining implements consisted of a butcher's knife, a tin pan and a crowbar.'

He talks of finding pounds of the stuff. I was sick with envy.

I sat down in a foot of water to cool my hot indignation. I knew I was not long for Downieville and this allowed me to consider the real reason for my mood. That morning I had been chatting with a local man over a cup of coffee and let slip the reason for my presence there, hoping inwardly that he might be impressed. He was no such thing.

'Wasting your time diggin' here,' he said. 'Best way to find gold on this river is with a wet suit and a face mask. Just float downstream and search all the crevices in the rocks below the water. New gold gets washed down in the floods each year. Gets trapped and waits to be picked up.'

Why had no one told me this before? You jumped into a lovely river, gently flowing through achingly beautiful Californian scenery in temperatures hovering around a hundred degrees and floated along picking up gold. Now, that's what I call work.

I left the dregs of my coffee and sprinted twenty yards to the town's hardware store. Inside, I found the mining section and there, to my astonishment, was a face mask and a snorkel. This, I decided, was a sign from God. Who would have expected to find implements for snorkelling in a store located so high in the hills, so far from the sea? I quickly paid for the items and ran to my car with the thought that this was the perfect way to jump claims without being caught. Not only could I act dumb, I could also pretend to be a swimmer and not a miner. I could even feign drowning if the need arose.

I drove east out of town – the water was too shallow and fast to the west – and found a stretch that was calm and clear and probably stuffed to the gills with gold that could be harvested easily and in the manner of a gentleman of leisure. I stripped to my shorts, put on my snorkel and face mask and swam to the middle of the Yuba. Farther upstream, I sighted a large and craggy rock that would surely have pieces of gold trapped in the unevenness of its surface.

A heron landed on the bank and watched as I dipped my head under the water, allowing me to fully appreciate the pristine nature of the riverbed and the water that flowed over it. A small snake wound across the surface, not two feet from me while a rainbow trout, opalescent in the sunshine, watched it from below. I floated in perfect peace, a persistent mantra the only thing in the world I could muster, involuntarily, to shatter it. 'Gold, gold, gold!' it went.

I brought myself into the moment and swam against the current in the direction of the rock. Breathing easily through the snorkel, I manoeuvred into its lee and pressed my goggles against the rock's gold-filled creases. Mine, I was thinking, all mine.

And that was when I remembered that I needed reading glasses. At this distance I couldn't tell the difference between a golden nugget and a chicken nugget.

It was time to leave Downieville and I knew exactly where I was going next: to the place that once made John C. Fremont, Pathfinder of the West, one of America's richest men.

18

Nuggets and Wage Slaves

If all those who rushed to the goldfields had found nothing, then their disappointment would have been easier to bear. Nearly all of them found something, but most pulled out only what was needed to pay for provisions, equipment and bawdy entertainment. Some – the wisest, perhaps – quickly realized that riches were unlikely to come their way and went home.

The miners who remained probably did so because of the lucky ones; they were sustained and encouraged by news of extraordinary gold strikes. Like me, they kept hearing about *the guy* – the guy who pulled out a nugget *this big,* and then a hundred more like it; unlike me, they would find verification easy because *the guy* was on the claim next to theirs or half a mile downstream or in the makeshift saloon buying endless rounds of champagne and whiskey.

With a lack of geological knowledge, early prospectors relied on luck. 'Gold lies whar you find it,' they would say. For many of them there was no rhyme or reason for success. It either came to you or it didn't. Tales abounded of groups of men striking it rich on one claim while the next, thirty feet away, produced nothing at all. One Frenchman out hunting is reputed to have fired at a quail and missed, his bullet hitting a rock, which shattered to reveal a rich vein of gold. A freed slave who chose to stay with his 'master' dreamed that the floor of their cabin was rich with gold; they tore it up and found twenty thousand dollars' worth in the soil – worth six hundred thousand today.

One woman who had borrowed fifty dollars to set up a restaurant in Downieville was doing very well charging three dollars a meal to miners who were only too happy to enjoy her cooking. But she closed the kitchen on finding a large gold nugget embedded in the restaurant floor; she hired some men to dig it up and left thirty thousand dollars richer – a sum equivalent to nine hundred thousand today.

Of course, there were random finds, too, nuggets of the most extraordinary size and value. A Mexican pulling his cart down the street stopped to move a rock out of his way; it proved to be thirty-five pounds of solid gold. A nugget weighing in at seventy pounds was found in Wood's Creek, Tuolumne County, in 1848. And in 1858, a man walking along a well-used trail near Columbia stopped in his tracks, blinked, and went to examine an odd-looking boulder that had been casually ignored by everyone else who had ever used the path. It turned out to be a fifty-pound piece of gold.

The big daddy of them all, however, was the Calaveras Nugget, unearthed in 1854; it weighed 162 pounds.

Many just plugged away in small teams, using pans and long toms – sluices like wooden gutters ten or twenty feet long – and slowly amassed decent amounts. Occasionally, they would enjoy a red letter day. Horace Snow, a young man from Bridgewater, Massachusetts, steadily worked a claim at Agua Frio (now Agua Fria) near Mariposa, with his brother, Hiram, and after two years returned home a reasonably wealthy man.

Writing to a friend, he described a competition with his brother, Hiram, to see who could retrieve the most gold from a single pan; until 12 August 1854, their record had stood at five dollars' worth – a sum, he points out, that was the same as the monthly wage he had been earning only a few years earlier. Then, on that day in August, Hiram 'struck a place which looked nearly yellow with gold!! He returned in a few moments and how much do you guess he obtained? Only nineteen dollars seventy-five!' enthused Horace.

Horace thought about conceding to his brother – surely he could not trump such an amount – but he carried on. 'Our competition had become very great, so much so that my brother had to go and see me wash [my pan] for fear that I should come some game upon him,' he wrote. 'The pan when washed, the gold dried and weighed and . . . how much do you reckon I had? Only the pitiable sum of forty-eight dollars and fifty cents! Wasn't I tickled! Just think how exciting it must be to take a common tin pan such as dairy women use for milk and fill it full of dirt and wash out forty-eight dollars fifty in it?'

Gold in 1854 fetched about nineteen dollars to the ounce, which means that Horace's single pan contained a fraction over two and a half ounces.

For most Argonauts, the experience was not so good and they were bewildered that more gold-seekers kept coming. It seemed to them that newspapers back east reported the big finds but made no mention of the thousands of miners who were struggling to make a living in what they called the Great California Lottery.

William Swain had worked hard through the winter of 1849–50 and, with his friends, had built a cabin and moved tons of earth to dam a stretch of the South Fork of the Feather River in order to expose its gravel bed. After suffering from poor health, working in freezing conditions and icy water, he sent a letter to his wife, Sabrina, in which he wrote, crushingly:

'We finished the job on the Feather River and tested it, although under great disadvantages. I am satisfied that it will not pay to work it out.'

Choosing his words carefully, he went about managing her expectations, both of the gold he hoped to find and of the time it would take. His target when he set out from home was to return with ten thousand dollars; he would have to stay much longer if he was to get even close.

'You have kindly asked in your last letter if my expectations are being realized,' he wrote. 'My specific answer to your kind question is that my expectations are not realized. We have been unlucky – or rather, by being inexperienced, we selected a poor spot for a location and staked all on it, and it has proved worth nothing.

'Had it proved as it was expected when we took it up, I should have more than realized my most sanguine expectation, and I should have today been on my way to the bosom of my family in possession of sufficient means to have made them and me comfortable through life. But it is otherwise ordered; and I mostly regret the necessity of staying here longer.'

By summer 1850, the reality was that most of the easy gold had been picked up and the fruitful sections of rivers had been claimed. One miner, on hearing that the second year of the rush was in full flush, wrote home: 'The fact is, we are too late in this country to make a fortune and every season makes the matter worse. There are now no new rivers to be found in California. All have been explored from their mouths to their sources, from the Mariposa in the south to the Trinity in the north.'

To his mother, William wrote of his fellow miners: 'Nine-tenths of them are sick at heart! Aye! Downhearted and discouraged! And many of them have great reason to be dis-heartened. Thousands who one month ago felt certain that their chances were sure for a fortune are at this time without money or any chance of any and [are] hundreds of dollars in debt. Certainly such a turn of fortune is enough to sicken the heart of any man.'

It was not unusual for miners to write home, urging their loved ones to spread the news that to travel to the goldfields was now folly. But such warnings served only to convince would-be Argonauts that the miners in California were growing rich and wanted to keep all the gold for themselves.

Where lucky prospectors did find riches, they often squandered them. During a period spent with miners at Mokelumne Hill in Calaveras County – where the placers were so rich that miners were restricted to claims measuring just four feet by four – Bayard Taylor wrote: 'It would have been an interesting study for a philosopher to note the different effects which sudden enrichment produced upon different persons, especially those whose lives had previously been passed in the midst of poverty and privation . . . It was not precisely the development of new qualities in the man, but the exhibition of changes and contrasts of character, unexpected and almost unaccountable.'

Those who were not used to manual labour tended to be disappointed with the ounce or half-ounce a day that they made and would hoard it, he said. However, those used to tilling the soil or performing back-breaking work received their rewards joyously – and set about spending them as fast as they could. If you were used to digging for potatoes but now the same labours resulted in pieces of gold, then that was indeed reason to celebrate.

'It was no unusual thing to see a company of these men, who had never before had a thought of luxury beyond a good beefsteak and a glass of whiskey, drinking their champagne at ten dollars a bottle and eating their tongue and sardines, or warming in the smoky camp-kettle their tin canisters of turtle soup and lobster salad,' observed Taylor.

Superstition was rife. On one occasion, Taylor bumped into

an Englishman, a former private in the cavalry, who was enjoying a considerable amount of success and was convinced it was due to his refusal to work on a Sunday. It was the Sabbath when he arrived at the diggings and he was hungry, but he resisted pressure to labour in return for bread.

'I'll dig o' Sundays for no man,' he told Taylor, in an accent that might explain the provenance of Dick Van Dyke's in *Mary Poppins*. 'An' I didn't, an' I had a hungry belly, too. But o' Monday I dug nineteen dollars, an o' Tuesday twenty-three, an o' Friday two hundred an' eighty-two dollars in one lump as big as ye fist, an all for not workin' o' Sundays!'

The miner then bought dozens of bottles of champagne, beer and brandy for everyone at his camp.

By 1850, as unclaimed land became rare and pickings grew slim, tensions built up on ethnic grounds, particularly in the southern mines where the presence of Mexicans – with whom America had only recently been at war – was resented. Within two years of James Marshall's gold find, one-quarter of the population of California were foreigners, and American-born miners began to take exception to them. Robbery and murder on both sides became commonplace; the days of low crime had gone.

When the easy placer gold dried up, particularly in the northern goldfields, ever more elaborate dams and wooden flumes like broad gutters or narrow aqueducts were constructed to divert rivers from their paths and give the miners more and longer access to river beds. The dams grew higher and wider the flumes sometimes as long as ten or twenty miles, and they were terribly inefficient and uneconomical. Teams of men would often take months to dig the dams and erect the flumes, only to find that the riverbed they were trying to access bore little gold. And then the dams and flumes would be washed away come the high and destructive floods of autumn and spring.

However, the expertise gained in moving water was put to practical use once miners realized that ancient river beds – long dried up when their flows were diverted by natural disaster or geological erosion – could be rich in gold too. The only problem was that the bedrock underpinning these was often dozens or hundreds of feet below them, and removing the topsoil amounted to a huge undertaking.

In 1853, a Frenchman, Antony Chabot, and an American, Edward Matteson, began using high-pressure hoses to wash away soil; once Matteson added a nozzle, the pressure it created rained down environmental destruction on a massive scale. This was the hydraulicking that Terry McClure had described to me. (The Romans had used water systems too during their plunder of Spain's gold, but the addition of the Matteson nozzle made Californian hydraulicking far more potent.)

Entire mountainsides were washed away, rivers became blocked with boulders and slurry, and farmland and pasture were ruined by avalanches of mud. While some companies busied themselves in taking apart the landscape, others began to dig directly into it.

By October 1850, placer miners in Grass Valley, in what is today Nevada County, were disappointed that their claims seemed to have played out and they enviously eyed the continuing success of the nearby camps, Rough and Ready, and Deer Creek, which would later be called Nevada City. Thirty-foot square claims in Grass Valley that had once changed hands for thousands of dollars were now selling for fifty.

One evening, a man called George McKnight, who had been struggling to earn a living in Grass Valley, climbed a hillside to gather firewood. According to the writer Robert Welles Ritchie, in his colourful 1928 book *The Hell-Roarin' Forty-Niners*:

[McKnight] carelessly kicked at a low ledge of rock just showing above the pine needles. A fragment crumbled away from the toe of his boot. Some God-given spur of curiosity prompted this fellow McKnight to stoop and examine this chunk.

The cleavage surface showed white as coconut meat, and through the glassy crystalline substance ran a ribbon of rich yellow, all clotted like honey in the comb. The wood-gatherer took his piece of rock back to his cabin and pounded it to a powder in the bottom of an iron kettle. He washed that coarse powder in a gold pan. The white powder slopped away over the pan's edge with the dribblings of water leaving – pure gold!

McKnight immediately staked a thirty-foot claim, named the location Gold Hill and set about digging deeper and deeper into the ground. By 1857 Gold Hill Mine had produced four million dollars' worth of gold, an amount equivalent to one hundred and twenty million dollars today.

This form of mining was known as lode mining as it involved digging into the ground in pursuit of golden fronds from the Mother Lode. The miners were confused over how gold could find its way into rocks, but they at least began to understand that, with erosion, this is where the placer gold they had been finding had originated. Their biggest concern was in figuring out a way to separate the precious metal from the rock. McKnight's first attempts at beating the quartz into powder inside his kettle were scaled up and machines called stamp mills were introduced. These consisted of a row of five or more vertical wooden logs that were thrust up and down on a rotating axle to smash the gold-bearing quartz into powder. The contraption was usually powered by a waterwheel, in Gold Hill's case this was fed from nearby Wolf Creek.

The end result was a sort of slurry that contained pulverized gold and pulverized quartz. When mercury was introduced to the equation, the gold stuck to it – gold has a natural affinity for mercury – and the quartz powder was washed away. The gold was then separated from the mercury through heating, a process that gave off poisonous fumes and, again, did the environment no good at all. Gold prospectors and anglers still regularly find mercury in the otherwise pristine waterways of California.

Of course, hydraulicking and lode mining were not enterprises undertaken by one or two men. To carry them out properly required investment – and lots of it. So with rivers claimed or emptied and easy pickings gone, many of the 49ers found that the only way they could support themselves was to take a job with the mining companies that embarked upon these costlier industrial-scale operations.

They had become wage slaves and were no better off than the day they had left home, California-bound and filled with hope.

19

The Luckiest Man in California

I was sitting on a stool at the bar of the Hideout Saloon in Mariposa, drinking whiskey and feeling very pleased with myself. I had just partaken of a truly golden evening and, though light in the pocket, I was replete in the belly and satisfied that I had spent the past few hours if not wisely, then memorably.

Earlier in the day, I had driven 250 miles south from Downieville. Like a 49er come to town after months working a claim, I had checked into a room at the Mother Lode Lodge (with a bed!) and then ducked (inevitably) into the Gold Coin Bar where, after downing several Grand Gold Cosmo cocktails and an indeterminate number of Jose Cuervo Gold tequila shots, I had repaired to a table to enjoy the 'hungry prospector meal' (fourteen-ounce rib-eye steak). Catching the eye of a passing waiter, I requested a bottle of Fosters Gold lager with which to wash it all down. The bar didn't sell it, but you can't have everything.

From there, I teetered along Bullion Street and ducked in and out of the splendidly old and well-preserved corners of what was undoubtedly one of the Gold Rush's most important seats of law, wealth and government. California's oldest courthouse is situated here, and many disputes over gold claims and ownership were settled in its beautiful first-floor courtroom (which is still in operation today). The town is steeped in mining history and peppered with beautiful gold-era buildings.

I had come here because I thought it might bestow on me some luck, and I needed it. I had begun to form a fascination with John C. Fremont, the explorer whose guides had led tens of thousands of Argonauts to California, not least Sarah Royce and William Swain, and I figured that if one-hundred-millionth of his luck rubbed off on me, then I might be happy, not to mention rich.

Fremont was one of the greatest heroes of the United States in the nineteenth century. Born in Savannah, Georgia, in 1813, to a French adventurer and the woman to whom he was bigamously married, Fremont possessed uncommon reserves of curiosity and drive, recklessness and determination, foolishness and verve. A colonel in the army, he was handsome, smart – if not wise – and hungry for fame, a fame that eventually came to him with a degree of infamy.

During the 1840s, explorers were feted, and none more so than Fremont, who undertook four ground-breaking expeditions, opening up the Oregon Trail and routes that would facilitate expansion west and the fulfilment of Manifest Destiny, the expansion of the United States from coast to coast. The books he wrote about his journeys were devoured by the American public, who endowed upon him his soubriquet, Pathfinder of the West. Perhaps because of the ignominious nature of his birth, Fremont was desperate for success and recognition, and this often led to rashness bordering on stupidity.

As an army officer he had a tendency to ignore or exceed his orders, constantly searching for ways to drape himself in glory. The worst examples of this involved, in 1846, attempts to provoke war with Mexico through unwarranted incursions into California at the completion of his third expedition, ostensibly to establish a route to the Pacific coast. At one point, ordered by the Mexicans to leave the territory, he found an excuse to attack a Native American village under Mexican protection, killing as many as 170 people. On another occasion, he and his band of fifty or so men encountered three well-connected Mexican Californians whom Fremont ordered to be put to death, arguing that he had 'no room for prisoners'. (These events came to haunt him when he later ran, unsuccessfully, as an anti-slavery Republican candidate for the office of president. No one wanted a murderer for a president.)

The Mexicans did not take the bait so Fremont first encouraged, then took over the leadership of a rebellion among American settlers in California that came to be known as the Bear Flag Revolt. It lasted only twenty-six days but it allowed Fremont to take Sutter's Fort and to attack and imprison senior Mexican officials. Fortunately for him, war with Mexico had already broken out on the Texan border – if it had not, he would have been in serious trouble with his superiors.

As it was, Fremont was made military governor of California, a position he refused to give up the following year when ordered to during a bout of internecine jostling between the War Department and the Navy Department. He was charged with mutiny, disobedience and conduct prejudicial to good order and discipline, charges of which he was found guilty. President Polk commuted Fremont's sentence – not least because it was generally agreed that the country owed him a debt of gratitude – and ordered him to return to duty. But the headstrong officer refused, resigning his commission instead.

Next, he did two things: first, he bought – unseen – a forty-four-thousand-acre parcel of land called Rancho Las Mariposas in the southern foothills of the Sierra Nevada Mountains, impulsiveness being one of his defining traits. Second, he began to plan an expedition, his fourth, to find a route for a railroad that would link St Louis with San Francisco. The expedition, in the winter of 1848, was a disaster and ten of his party of thirty-five died in freezing alpine conditions. Although frostbitten, Fremont survived.

While he and his party were struggling through the mountains, his wife, Jessie, daughter of Senator Thomas Hart Benton, the original proponent of Manifest Destiny (and a man with the terrific nickname of 'Old Bullion'), was travelling to meet him via the Panama route to San Francisco. Jessie was fiery and clever, possessing the wisdom that Fremont lacked, and it would not be an exaggeration to describe the couple as the Elizabeth Taylor and Richard Burton of their day. Everyone loved them; tales of her beauty and wit, and his derring-do, preceded them wherever they went.

On his way to meet Jessie in San Francisco, Fremont bumped into a large group of Mexicans from the border town of Sonora and – completely unaware that gold had been discovered while he was in the mountains – he asked them where they were going. 'Why, to dig for gold, of course,' came the reply. There and then, with hardly an introduction, Fremont employed a couple of dozen of them to prospect for gold on the land he had bought, land that he had still not seen – the deal being that they would split whatever they found, half to the Sonorans, half to Fremont. They agreed.

After a grand reunion, Jessie and John set up home in Monterey, while the Sonorans, true to their word, sent them fifty per cent of everything they found. It turned out that Fremont, The Pathfinder, had bought the richest parcel of land on the Mother Lode. The first shipment was worth eighteen thousand dollars – about half a million today. And it just kept coming. Jessie wrote of receiving 'buckskin bags filled with gold dust and lumps of gold.'

After a year, the Sonorans decided they had enough money and sent word that they planned to return home. Staggeringly, Fremont decided not to travel to Mariposa himself for a final reckoning, choosing instead to send a messenger to check that the Mexicans left with no more than their fair share. According to Jessie: 'This they did with scrupulous honour, not taking an ounce more than their stipulated portion.'

So, here I was in Mariposa, to find my cut.

I ordered a beer from Brenda, the bartender. She caught me looking at the stone wall that ran the length of the bar, and which had coins embedded in it. 'They're original dry stone walls built by the Chinese in the Gold Rush,' she said. I poked my finger into a gap between two rocks. 'There isn't any cement at all, just good craftsmanship. It was customary to put a coin or two in the spaces when you were doing well, and if a miner hit on hard times he would come and take a few out. If his luck improved, he'd put them back and maybe a couple more if he could afford it. It kept people going when times were tough.'

A man sitting at the bar next to me listened to this story and looked at the coins earnestly. He had been hugging the same empty bottle since I came in.

'Buy you a beer?' I asked.

'Sure,' he said. 'Thanks.'

His name was Matt and he was in his forties. His hair and beard were white and his face was tanned from working out-doors. He had spent most of his life in construction and forestry but cutbacks had taken their toll. He was unemployed.

'Used to get a lot of work at Yosemite, but there isn't any right now,' he said, cradling his drink. 'Hasn't been for some time. And because of the recession there's no building. If you can't get work in forestry or construction, there isn't a lot you can do in this town. Right now I'm in recycling.'

By recycling, Matt meant collecting discarded bottles and aluminium cans, some from the trash, some from bars and restaurants. There was a recycling centre in town that paid by weight for the cans, while the bottles were returned for a small deposit on each. It was hard work that brought in an average of fifty or sixty dollars a day, he said. Hundred on a good day.

I asked if he ever supplemented his income by prospecting for gold. 'No, but some people do,' he said. 'You still hear of people pulling nuggets out of Mariposa Creek. You need to meet my friend Shannon. He knows where to find gold.'

Perhaps this was it, the piece of luck I had needed, I thought to myself. Perhaps Shannon was *the guy*.

We talked and drank some more before I decided to turn in. I would meet Matt and Shannon at 7am the next morning outside the Sugar Pine Cafe downtown. I crawled between crisp white sheets at the Mother Lode Lodge and emitted loud sighs, thinking before I fell asleep that I need not worry about bears, ticks, poison oak, mountain lions, rattlesnakes or bubonic plague.

At least, not until tomorrow.

Mariposa is the Spanish word for butterfly and the area is famed for them, though I couldn't see any floating on the breeze at that time of the morning. It was another beautiful day, chilly outside the Sugar Pine Cafe but destined for reliable Californian calescence later. Matt had been a little late and

Shannon hadn't shown up at all, but it didn't matter. The town was sleepy and rushing it seemed wrong, if not cruel. We sat at a table in the open air; this was where the men met each morning, bringing coffee with them from the cheaper gas station down the road. The people in the cafe seemed to understand and made no moves to raise the issue in any proprietorial way, but I bought two more coffees from inside anyway.

Matt was dressed in the same pink T-shirt and cut-off jeans from the night before and he seemed a little bedraggled. I felt depression might have been knocking at his door but I also figured that turning up at this hour every day was not indicative of someone who had given up. I told him I felt it would be only fair if I made up any recycling losses he and Shannon incurred while they showed me where to prospect. Matt said that was between me and Shannon, if we ever found him; Matt wouldn't accept anything. He was obviously not afraid of hard work – he just couldn't get any. All over America, all over the world, there were people like Matt. I have no idea whether he considered this kind of 'recycling' to be demeaning, but he wasn't ashamed to tell me he did it. It was work, and that gave him dignity.

After an hour Matt suggested we get up and look around town for Shannon. We found him leaning into a trashcan, a trolley next to him filled with tins and bottles, on the main road that ran through Mariposa. He was a lean forty-seven-year-old dressed in a dirty grey T-shirt and jeans covered in dried mud. He had a salt-and-pepper beard and a shock of brown hair that was greying behind his ears.

Matt introduced us and after some initial reluctance – he was having a good morning – Shannon offered to take me to his spot. Matt would find a safe place for the cans and then join us. I imagined a long drive farther into the pine-clad mountains and a walk, blindfolded perhaps, the last mile or so to Shannon's secret gold-bearing place, but after just two minutes he directed me to a side road and told me to park outside a row of bungalows. We ducked into some bushes and he led me down a low bank to a creek that was barely flowing. There, in the bed of what he assured me was a violently flowing river during winter and spring, were his diggings.

This was Mariposa Creek and it was on land originally owned by The Pathfinder of the West.

There was a bucket of concentrates left over from the previous day and Shannon scooped some out and began panning. There was hardly enough water in the river bed to do the job.

'I'm originally from Detroit,' he said. He had few teeth but that didn't stop him smiling when he talked, gently rocking his pan. 'I used to work with wood in saw mills and pallet factories, and with fibreglass, moulding seats for McDonald's restaurants. I worked in Louisiana too – got my merchant seaman's certificate on shrimp boats. Then I went back to Detroit but there was no work, so I came out here in 2008 when gold was around a thousand dollars an ounce. I can still remember the first time I found some gold – it was in the Merced River at Briceburg. Yeah! You never forget that. I kept prospecting and never went back to Detroit, but I didn't strike it rich. Gold went close to two thousand dollars an ounce a while back, but these days you're lucky if you get thirteen hundred for it.'

Matt caught us up. He pushed his way through the bushes and walked along the riverbed, nodded and sat on a rock. He lit a cigarette and took a can of beer from a brown paper bag. He offered us one but we both declined.

Some years earlier, Shannon had been hit by a car in Florida and the third and eighth vertebrae in his spine had been crushed. 'I need surgery but they won't give it to me, so I wait around for disability and I pan for gold,' he said. 'But I can only do it three or four hours a day before it gets too painful. Gold, disability and recycling keep me going.'

I didn't know who he meant by 'they' because I assumed he

did not have medical insurance, but it was clear that his situation rankled. It was also clear that he, too, was not afraid of hard work; he just couldn't do it all day.

After some time, Shannon showed me the single flake of gold he had panned from the previous day's bucket of concentrates. He was doing no better than I was. I looked at his spot and decided to pass on it. I didn't think it would ever bear good gold but I didn't say so. The fact that I was becoming more discriminating was, for a moment, a source of some satisfaction; a month ago I wouldn't have had a view on the matter.

I thanked the men, said goodbye and drove fifteen miles north to Briceburg to try my luck in the Merced River, where Shannon had found his first gold. John Borthwick spent a couple of months in that area but, frustratingly, in *Three Years in California* he provides no account of exactly where he stayed, how he spent his time or how much gold he found. I suspected none, but that didn't trouble me. I remained convinced that I would soon hit a pay streak.

There was a public panning area, which I shunned, instead walking upriver a short distance. My chosen spot was sheltered by high banks on either side, covered in scree and low shrubs. The river was very shallow and stream-like in places; I walked into it in my boots without fear of water swilling down my ankles. I would just prospect at first to see if the spot was worth working, so I had only my pan and a trowel with which to scoop up a few samples.

I filtered out the sounds of water. The silence that remained was humbling and for a time I sat on a rock and absorbed it, annoyed that the tinnitus echoes of the morning would not quite abate inside my head. If I hadn't had a rock on which to sit, I suspect I would have slumped into the water or fallen from giddiness.

I dug behind a rock above the waterline with my trowel and found a smattering of gold flakes in my pan at the first wash, so I walked back to my car to fetch some buckets, my pick and shovel; this was promising. I spent the afternoon digging and then panned two buckets of concentrates, but the finding of colour at the first attempt was a fluke. I washed out only a few more flakes. I convinced myself that this was not a reversal; it was a test to see whether I had the resolve to go on, and if I did then I would be rewarded on some river at some moment by a crevice filled with nuggets or a grotto lined with quartz and gold.

But it did not take long for my optimism to fade away. I sat in the icy water and felt as disappointed as a 49er fooled into buying a spent claim. 'Don't give up,' I heard myself say. I would move on again in the footsteps of John Borthwick, this time heading sixty miles north to Sonora where my luck would change for the better.

Only the night before, someone in the Hideout Saloon had told me about a guy there who had just pulled out a nugget *this big*.

20

Sonora, E Clampus Vitus and the Plate of Brasse

A few days later, I travelled on through the old mining towns of Coulterville and Jamestown to Sonora. I was keen to see Sonora and to prospect there because John Borthwick had been most taken by it. While William Swain and Sarah Royce had remained in the northern and central mining districts, to which I was slowly returning, Borthwick had been keen to experience the southern mines, which seem to have been frequented by a higher percentage of foreigners, making the towns that serviced them more cosmopolitan and sometimes more raucous.

After roughing it for some time in mountain mining camps, Borthwick found Sonora – named by miners from Sonora, Mexico, in 1851 – most agreeable, and so did I. Much architecture survives from the post-gold era, along with more theatres, restaurants and bars than is warranted by a population of just seven thousand.

The populace had been higher (though largely itinerant) back then. As well as Mexicans, there were Americans, French, Chinese, Germans, Italians and other nationalities who were generically labelled 'Dutch'. In contrast with the many descriptions of the northern mines as being hard and often dour, with nothing but drunkenness and gambling to alleviate the boredom, Sonora was bright and cheerful, imaginative and fun. Borthwick noted a somnolent indifference to work among some of the southern miners, coupled with a level of unbridled energy when it came to enjoying themselves. There had been music and dancing, a variety of international cuisines to be enjoyed and, every Sunday, bull and bear fights to entice hundreds to the town's bull ring. There seems to have been a buzz about the place.

'One could live here in a way which seemed perfectly luxurious after cruising about the mountains among the small out-of-the-way camps,' wrote Borthwick. 'For, besides having a choice of good hotels, one could enjoy most of the comforts of life; even ice-creams and sherry cobblers were to be had, for snow was packed in on mules thirty or forty miles from the Sierra Nevada, and no one took even a cocktail without its being iced.'

The town had about it a certain panache that delighted Borthwick's artistic sense of beauty and colour; he had been particularly impressed with the effort the miners made with their appearance: 'On Sundays especially, when the town was thronged with miners, it was quite gay with the bright colours of the various costumes. There were numerous specimens of the genuine old miner to be met with – the miner of '49, whose pride it was to be clothed in rags and patches; but the prevailing fashion was to dress well. Indeed there was a degree of foppery about many of the swells, who were got up in a most gorgeous manner.

'Some men wore flowers, feathers or squirrels' tails in their hats; occasionally the beard was worn plaited and coiled up like a twist of tobacco or was divided into three tails hanging down to the waist. One man of original ideas, who had very long hair, brought it down on each side of the face and tied it in a large bow-knot under his chin.'

Unlike the northern mines, where women still formed only a tiny fraction of the population, in Sonora: 'The numbers of Mexican women with their white dresses and sparkling black eyes were by no means an unpleasing addition to the crowd.'

The placers around Sonora were exceedingly rich in the early days, so much so that the spending of gold became mind-bogglingly cavalier. According to the Gold Rush historian John Puttnam:

> The amount of gold mined near the town of Sonora was so great, and attracted so many men to the area, that no attempt to accurately weigh or measure it was made for normal day to day transactions.
>
> A pinch, or the amount that could be held between the thumb and forefinger, was called a dollar. A tea-spoon full passed for an ounce; a wine glass was worth a hundred dollars, and a tumbler a thousand dollars.

It was not uncommon for mining towns to burn down and be rebuilt in astonishingly short order. While Borthwick was in Sonora, he witnessed this. It was 1am on a June morning in 1852 when, looking out of the window of the French hotel in which he was lodging, the artist saw a house on fire a hundred yards away. He helped his landlady to clear out as much furniture as possible, but then had to run to a hillside where thousands more watched as Sonora was consumed in its entirety, save for two churches and a dance hall that was immediately requisitioned as a bar. Mercifully, only two people died in what came to be known as the Sonora Great Fire.

While there had been some gnashing of teeth over this event, most of the crowd had remained calm. They might have lost a few items of clothing, some furniture, perhaps, wrote Borthwick, but: 'The greater part of the people were individuals whose wealth was safe in their buckskin purses.'

As soon as the flames had died down, rebuilding began: 'The hot and smoky ground was alive with men clearing away rubbish; others were in the woods cutting down trees and getting out posts and brushwood, or procuring canvass and other supplies from the neighbouring camps. In the afternoon the Phoenix began to rise. Amid the crowds of workers on the long blackened tract of ground which had been the street, posts began here and there to spring up; presently, cross-pieces connected them and before one could look round the framework was filled in with brushwood.'

The speed with which business was resumed staggered Borthwick. Rough houses and hotels, stores and saloons were knocked up with uncomplaining zeal. Sign-painters did a roaring trade demonstrating that this or that establishment was open and trading: 'Everyone, as soon as lumber could be procured, set to work to build a better house than the one he had lost, and within a month Sonora was in all respects a finer town than it had been before the fire.'

While Borthwick admired the cosmopolitan nature of Sonora, many American miners resented it. In particular, they felt it unfair that foreigners and Mexicans, against whom they had so recently been fighting, could turn up and take gold from American soil. In an effort to drive these foreigners away, California's young legislature introduced the Foreign Miners Tax, which at twenty dollars per month (equal today to about six hundred dollars), served only to fuel growing tensions in the mining camps, as many of the foreigners refused to pay.

In a letter to his brother in May 1850, Bernard J. Reid, a miner living on Woods Creek, which flowed through Sonora, described the discord that the tax generated, and some of the loss of life that could be directly attributed to it:

Within 10 miles of Sonora there is a large marjority of foreigners. They thought the tax oppressive and were deliberating resistance. The ringleaders and fomenters of trouble were chiefly French Socialists – Red Republicans . . .

They collected the simple and ignorant Mexican and Chilean peasantry – harangued them with inflammatory speeches – denounced the tax as 'unjust' and 'tyrannical' – and said IF IT MUST BE PAID, pay it in powder and ball. Frenchmen from all quarters collected about Sonora [18 May] all armed to the teeth, and affairs grew serious.

The excitement was heightened by an American who rashly undertook to cheat a Chileno out of his hole – by representing himself the Collector and demanding the tax. The Chileno said he had not then enough money to pay it. 'Then you must quit work.' The Chileno quietly left the hole – but no sooner was he out than the perfidious scoundrel jumped into it in his place. This exasperated the Chileno who with one blow cut the jugular vein of the American and he died in a few minutes.

The Americans in and about Sonora felt apprehensive for their safety and sent runners to Mormon Gulch and Jamestown for reinforcements. That night about 400 Americans [went] there under arms – they remained next day which was the last 'day of grace' given by the Collector – and owing no doubt to the determined front they presented, no difficulty took place, except that a Mexican drew a knife on the Sheriff (newly elected) but was killed on the spot by the Sheriff's deputy. One Frenchman was arrested for inciting the mob by inflammatory harangues.

Thus ended, or rather was prevented, the war which if begun would have raged furiously in this quarter. It would have become a war of extermination, for the want of any power on either side to arrest it, and it would have swept to its end so quickly. For several days a painful suspense prevailed, and we were all in readiness to march at a moment's warning to the assistance of the authorities and our fellow citizens in danger. Now all is quiet and many of the foreigners have left. I am sorry to say that several Americans who reap a rich harvest of profit off the foreigners, encouraged the disaffection, denounced the law, and preached actual treason.

The tax was repealed the following year and replaced with another in 1852 that specifically targeted Chinese miners, requiring them to pay three dollars per month. Because it was customary for the Chinese to be run off any claims that paid well, they were left with the dregs of those that were unwanted and most earned as little as six dollars per month. This tax was not done away with until 1870.

I did some prospecting around Sonora and the nearby (wonderfully preserved) Columbia and found some gold, but not much. One night, however, I enjoyed good fortune of a different kind. I was sitting in a bar on Washington Street that permitted smoking indoors – in health-conscious California. This was a type of establishment arguably rarer than gold itself and I wanted to immerse myself in foul-smelling clouds of billowing blue smoke while enjoying a cold beer, just for old time's sake. After all, this was the kind of atmosphere to which the 49ers would have been often exposed. They were prolific smokers.

The establishment was called the Zane Iron Horse Lounge and the bartender told me he was a partner in the business; so were the other three bartenders, and that is how they had circumvented California's strict anti-smoking laws. The laws were aimed at protecting employees; if you didn't have any, then everyone else in the place was there of their own volition, not because they had to earn a crust in a smoky, unhealthy environment.

'Neat, huh?' A large smiling man at the stool next to mine leaned forwards, took a drag on his cigarette and offered his hand. 'Forrest Albright,' he said.

I introduced myself and Forrest asked me what had brought me to Sonora. When I explained, he sat upright on his stool.

'You're prospecting?' he asked. 'You heard of E Clampus Vitus?'

I had, indeed, heard of E Clampus Vitus, but it had remained something of a mystery to me. Almost everywhere I went where gold-era buildings still stood I had seen plaques on the front of them explaining their significance – for example, I first became aware of 'Juanita' from one of these in Downieville on the building where she had been 'tried'. They performed a most valuable service and reminded me of the blue plaques on buildings in the UK that were once occupied by distinguished people. However, I was always slightly nonplussed by the concluding proclamation that the plaque in question had invariably been put there by E Clampus Vitus, something that the dustier recesses of my brain erroneously told me was a disease not dissimilar to tetanus. (The name actually means nothing, either in Latin or any other language.)

It is, as Forrest went on to explain, an organization that's dedicated to charitable causes, preserving the history of the Mother Lode region – and having a good time. Most often today, E Clampus Vitus says of itself that it cannot decide whether it is 'a historical drinking society' or 'a drinking historical society'. Another definition it applies to itself is: 'A wart on the backside of the American tradition of Fraternal Brotherhood.' It was probably established by miners who either didn't want to be members of organizations such as the Freemasons, or who couldn't get into them.

Deliciously, the most senior member of each chapter was called the Noble Grand Humbug and the organization's constitution declared that 'all members are officers and all offices are of equal indignity.' It transpired that Forrest, a particularly cheerful and endearing character, was the Noble Grand Humbug of the Matuca chapter, which covered the counties of Mariposa, Tuolumne and Calaveras. I cannot tell you how delighted I was to hear this.

I told Forrest about all the E Clampus Vitus plaques I had read, and found myself thanking him and his comrades for making my journey all the richer. In turn, he impressed upon me the spirit in which the organization had been founded and continued to thrive, reminding me that its most important role in the days of the Gold Rush was to take care of the 'orphans and widders' of miners who had come to a sorry end.

As we drank and took in the secondary smoke, Forrest told me a very funny story.

In 1579, when Sir Francis Drake landed for a while in northern California, he is said to have had a brass plaque forged and nailed to a tree, claiming the territory for Queen Elizabeth I. The existence of this plaque is known because one of Drake's party, Francis Petty, wrote of leaving behind a 'plate of brasse' bearing this inscription:

SONORA, E CLAMPUS VITUS AND THE PLATE OF BRASSE

BEE IT KNOWNE VNTO ALL MEN BY THESE PRESENTS.
IVNE.17.1579

BY THE GRACE OF GOD AND IN THE NAME OF HERR
MAIESTY QVEEN ELIZABETH OF ENGLAND AND HERR
SVCCESSORS FOREVER, I TAKE POSSESSION OF THIS
KINGDOME WHOSE KING AND PEOPLE FREELY RESIGNE
THEIR RIGHT AND TITLE IN THE WHOLE LAND VNTO HERR
MAIESTIEES KEEPEING. NOW NAMED BY ME AN TO BEE
KNOWNE VNTO ALL MEN AS NOVA ALBION.

G. FRANCIS DRAKE

American historians had prayed for the day when the plaque might turn up. After all, it marked a significant moment in the nation's history. One of the most distinguished of these was Professor Herbert Bolton, a lecturer in Californian history and director of the Bancroft Library at the University of California, Berkeley. It might fairly be argued that Professor Bolton was a little obsessed with finding the plaque, having urged hundreds of his students over the years to contact him on hearing even the slightest whiff of information about its existence.

Unfortunately for Professor Bolton, he was a member of E Clampus Vitus, and one of the main tenets of the organization requires members to perpetrate pranks on other members. Knowing of his interest in the plaque, several of them set about making one in a way that, they assumed, would be exposed as a hoax. This was in 1933.

The fake plaque was deliberately left in Drake's Bay, about thirty miles north-west of San Francisco, where the explorer is assumed to have landed. A chauffeur did, indeed, find it while waiting for a quail-hunting client, but several weeks later he threw it out of his car window near San Quentin Prison. It was three years before someone else, a pheasant hunter, found it again. This person had a friend who had studied under the professor. After a cursory examination of the plate, and with the backing of seventeen members of the California Historical Society, the academic bought the fake for the equivalent of about sixty-thousand dollars.

The Clampers were horrified – it had never been their intention to call into question the reputation of a fellow member – but the hoax had grown out of control and they did not feel they could tell Professor Bolton about it directly; he might use the brass plate to beat them to death. Instead, they attempted to tip him off. Several weeks after the 'discovery', they made another fake plate to demonstrate how easily one could be turned out by, *ahem*, would-be mischief-makers; the warning fell on deaf ears. A member of the order wrote a letter to the professor from the non-existent 'Consolidated Brasse and Novelty Company' offering 'special' brass plates for every occasion; but Bolton failed to take the hint. And, finally, E Clampus Vitus produced a pamphlet entitled '*Ye Preposterous Booke of Brasse*', actually explaining how and why the plaque was a fake, even stating: 'We should now re-claim [the plate] as the rightful property of our ancient Order.'

But Professor Bolton's faith – and that of his university – was unshakeable, even when other academics pointed out apparent inconsistencies in the plate's make-up and grammar. He went to the grave in 1953 still believing in its authenticity.

It was the 1970s before a series of tests by the Massachusetts Institute of Technology, the Research Laboratory for Archaeology and the History of Art at Oxford University, and the University of California's own Lawrence Berkeley Laboratory finally exposed the plate as a dud. During the intervening period, when it was thought to be genuine, copies had been presented to visiting dignitaries from around the world, including Queen Elizabeth II.

In 2003, after eleven years of research, a group of academics produced a report on exactly who was responsible for the hoax. It had been a few E Clampus Vitus members and history buffs who were long dead, their awful secret having been kept from Professor Bolton. At the press conference held to unveil the report's findings, a group of 'Clampers' were invited in order to add a little colour. One of them, Rick 'Cap'n Crunch' Saber, expressed regret over the fact that there were no minutes from E Clampus Vitus that could have settled the matter decades earlier.

'Nobody was in any condition to record them,' he said.

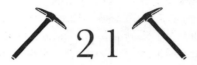

21

Stone Money, Paper Promises and the Gold Standard

I was lying in my tent near Sonora pondering the wonder of this thing that had me under its spell. It was nothing more than an element, no less natural than tungsten or arsenic, yet it had extraordinary power, a draw as irresistible as gravity. It was scarce, but its value was not in proportion to that scarcity; other factors – its colour, its timelessness, its density – took care of that. Yet what made a lump of gold better in my hand – by any system – than a rock? Would it have the same allure if it grew on trees?

Fifty years after James Marshall found gold, the German government paid Spain just over three million dollars for ownership of a tiny island called Yap, one of the Caroline Islands in Micronesia in the western Pacific Ocean. When German officials travelled there, keen to develop the island as a stepping stone for telegraphy services, they discovered two things: first, the Yapese did not like hard work and, second, they could not be persuaded to do it even with the inducement of money – whether paper, silver, bronze or gold.

The reason was that the Yapese had a currency of their own and it was made of stone, but not stone found on Yap or its near neighbours. No, this stone, a kind of limestone rich in white calcite, had to be transported some three hundred miles north-east from the island of Babelthuap in the Palau archipelago by brave men in flimsy canoes. Not only was this stone shinier and better to look at than anything found on Yap, but also the effort and danger involved in transporting it so far made it desirable. Size mattered, too, because the larger the stone, the more dangerous and difficult it had been to carry to Yap, and this made it rarer. Every ambitious chief and elder wanted one of these large stones, but only the bravest or most influential – those, perhaps, who could persuade young men to fetch them in return for favours, goods or animals – could acquire them.

Because a round trip of six hundred miles in a canoe to carry rocks was not something embarked upon lightly or regularly, these stones were, at any one time, finite, which meant that their value became quantifiable through the forces of supply and demand. They were carved into circles with holes through the centre for portability, ranged in diameter from two inches to twelve feet and were variously known as *rai* or *fei*. A *rai* that was three feet in diameter could be exchanged for a pig or a thousand coconuts; one measuring four feet across might atone for the theft of a woman; one measuring twelve feet across might buy you an entire village.

Of course, the sheer weight of the bigger, more valuable, *rai* meant that moving them around became impractical, so when they changed hands, the parties in the exchange would acknowledge the transfer of ownership and the stones would remain where they stood, at locations all over the island.

This left the Germans in fits of laughter. What fools these people were! The German officials decided to fine the recalcitrant Yapese until they could be persuaded to build roads, erect telegraph poles and so on. But if it was to be effective, what form should the fine take?

The American anthropologist, William Henry Furness II, visited Yap in 1903 to study the islanders and was told what happened next. In his subsequent book, *Island of Stone Money*, he wrote:

> At last, by a happy thought, the fine was exacted by sending a man to every [community] throughout the disobedient districts, where he simply marked a cross in black paint to show that the stones were claimed by the government.
>
> This instantly worked like a charm; the people, thus dolefully impoverished, turned to and repaired the highways to such good effect from one end of the island to the other, that they are now like park drives. Then the government dispatched its agents and erased the crosses. Presto! The fine was paid, the happy [owners] resumed possession of their capital stock and rolled in wealth.

The financial historian Peter Bernstein, in his book *The Power of Gold* (2000), describes the day in 1940 when, as a treat, his boss at the Federal Reserve Bank of New York, where he worked as a researcher, took him down to see the gold vaults five storeys below ground. 'The gold was stored in oversized closets, about ten feet wide, ten feet high, and eighteen feet deep,' recalls Bernstein. 'The closets were filled to the ceiling with towering piles of gold bricks, each brick the size of three large candy bars. The bricks weighed about thirty pounds apiece ... and were worth fourteen thousand dollars in those days, when gold was officially priced at thirty-five dollars an ounce.'

Even at that price, the amount of gold was worth – in 1940 – two billion dollars or, as Bernstein puts it: '[It] was a sum of money sufficient to buy four days' worth of the total production of goods and services in the United States.' Yet it was housed in a cramped room five storeys below the streets of New York City.

Most of this gold belonged to Britain, France and Switzerland and it was stamped accordingly on each bar, with a separate record of ownership kept in big bank ledgers. This was called 'earmarking' and it saved the governments of these countries the trouble of transporting gold when they had vast transactions to settle among themselves.

'If England lost gold to France,' Bernstein wrote, 'a guard at the Federal Reserve had merely to bring a dolly to England's closet, trundle the gold to the French closet, change the earmark and note the change on the bookkeeping records.'

Was this any less bonkers than anything that happened on Yap?

Furness described one family on Yap whose vast wealth went unquestioned because it was underwritten by a huge stone that no one had ever actually seen. Several generations earlier, a brave ancestor had travelled to Babelthuap, carved an enormous *rai* and sailed it on a raft towards Yap, only to lose it in a storm just short of the island.

'When they reached home, they all testified that the [*rai*] was of magnificent proportions and of extraordinary quality, and that it was lost through no fault of the owner,' wrote Furness. 'Thereupon it was universally conceded in their simple faith that the mere accident of its loss overboard was too trifling to mention, and that this few hundred feet of water offshore ought not to affect its marketable value, since it was all chipped out in proper form.'

The family's wealth was assured, even though the stone lay out of reach at the bottom of the ocean. It was just as useful to them as the unseen gold in the bowels of the Federal Reserve was to Britain, France or Switzerland.

In the century after the Spanish plundered Mexico and Peru, the quantity of gold sloshing around Europe increased by five hundred per cent while, at the same time, international trade grew apace. As merchants and traders began to travel farther and wider, carrying gold coins became both impractical and dangerous, a state of affairs that resulted in the creation of paper guarantees that promised to pay the bearer on demand a certain sum in gold. (Similar leather or paper guarantees had been employed in China as early as 118BC.) In Europe, these guarantees – known as running cash notes – were first issued by London goldsmiths into whose safekeeping the gold had been deposited.

Apart from the enormous exchanges between countries to settle imbalances in payments – the kind described by Peter Bernstein – hardly anyone settled debts with large amounts of gold once banks began printing paper money in the seventeenth and eighteenth centuries. However, in order to establish confidence in this paper, the world's advanced economies adopted the gold standard, the system by which they printed money only to the value of the gold they held.

To this day, British banknotes feature these words, signed by the incumbent governor of the Bank of England: 'I promise to pay the bearer on demand the sum of . . . [five pounds, ten pounds and so on].' Britain first established the gold standard in 1717 and finally dropped it in 1931 – and it is a popular myth that throughout this period you could go to the Bank of England and demand, say, ten pounds' worth of gold for your banknote. In fact, that was once true, but only up until 1914 when the government put an end to the practice in order to shore up gold reserves.

(It is worth noting here that when Sir Isaac Newton wasn't defining gravity or the laws of motion, which still govern space travel today, he served as warden and master of the Royal Mint, capturing counterfeiters and dreaming up the gold standard. Before this, silver had held sway. The British 'pound' originally got its name because it was worth one pound of silver.)

Some economists still go misty-eyed at the mention of the gold standard. The system was at its zenith between 1871 and 1914 when its proponents argue that it resulted in an unprecedented period of international trade, prosperity and peace; others say it was merely a symptom of the outbreak of such prosperity, which came about cyclically anyway.

The gold standard facilitated international trade because countries that adopted it set a fixed price for gold – it wasn't openly traded. And because a set amount of money – the money supply – was issued against the quantifiable amount of gold held in the country's vaults, at a fixed price, then the value of each country's currency could be precisely posited. So, for example, in 1834, the United States fixed the price of gold at 20.67 dollars an ounce, a price that remained fixed until 1933; in the UK the price was fixed at 4.24773 pounds sterling an ounce from 1821 to 1914. This meant that participants in international commerce could say for sure that one pound sterling was worth 4.866 dollars. Such certainty made for sure-footed trades, even when dates for making payment were in the future.

Internationally, the gold standard was self-correcting in a frustrating and emotionless see-saw kind of way. It could cripple countries when they had a little local economic difficulty. If, for example, you had a bad harvest and had to buy your wheat from abroad, you would end up with a balance-of-trade deficit and gold would leave your country. This meant that you could print less money (because you had less gold to back it up). The effect of this was that interest rates would rise, triggering a depression that would cause wages and prices to fall. This made your country's exports cheaper, everyone would want to buy them, your balance-of-trade deficit would disappear and your economy would recover.

Where booming countries were on the receiving end of gold, prices would go up and their exports would seem more expensive. Eventually, they would suffer a balance-of-trade deficit, their gold would flow out and their side of the see-saw would hit the floor. The idea was that neither periods of boom nor bust would last for long.

The gold standard saw to it that if your economy was in dire straits, you could borrow money but couldn't print your way out of trouble to pay your debts. It was suspended in 1914 when the First World War destroyed economies and shattered trust among nations who found themselves embattled but strangely reliant upon their enemy to play by the financial rules and not simply print cash to pay for munitions. The gold standard fell apart. It was rekindled in 1925 in a different form, whereby most nations could back their currencies using gold, dollar or pound reserves, but the USA and the UK could back their currencies only with gold.

Britain was forced to pull out of this model in 1931 during the Great Depression, when a run on the pound caused by a weakened economy saw reserves depleted as the pound was exchanged for the safety of gold. During the worst week of the run, forty million pounds' worth of gold (a figure equivalent to £2.4 billion in spending power today) was reclaimed from the Bank of England.

Many economists argue that this was a blessing in disguise; no longer constrained by the austere economic rules of the gold standard, the British government was able to devalue the pound and introduce low interest rates on borrowing, all of which stimulated the economy and brought about recovery.

Within a year of Britain abandoning the gold standard, most other countries – scores of them – had left it too, leaving only France, the United States, Belgium, Switzerland and the Netherlands still clinging to it. In 1933, with gold pouring out of the USA and into the vaults of the remaining gold standard nations – whose panic manifested itself in the desire to prop up their currencies with ever-greater quantities of gold – President Franklin D. Roosevelt came off the gold standard, nationalized gold and made it illegal for citizens to hoard it.

At about the same time as it became legal in America to be found in possession of whiskey, it became illegal to be found in possession of gold in any great quantity. The president signed Executive Order 6102 making it a crime for any individual, partnership, association or corporation to hoard 'gold coin, gold bullion and gold certificates within the continental United States.' It was not until 1974 that President Gerald Ford signed a bill that allowed Americans to 'purchase, hold, sell, or otherwise deal with gold' once again.

Order 6102 replenished the gold supplies lost to the panic of 1931. It made even more sense when Roosevelt oversaw a devaluation of the dollar, fixing the price of gold upwards at thirty-five dollars an ounce. This increased the value of the country's reserves, enabling the US Treasury to print money to that value and creating the inflationary pressures that, in conjunction with new fiscal programmes to promote growth, helped to haul the US economy out of the mire of the Great Depression. If you're not getting this, just think of it as economic smoke and mirrors: illusory but effective.

Towards the end of the Second World War, forty-four countries sent delegates to the town of Bretton Woods in New Hampshire to determine how the world's postwar economy would function. The upshot, with the USA now holding seven-tenths of all the world's reserves of gold, was that their currencies would be pegged to the US dollar, and the dollar would be convertible, on demand, into gold. This worked while the United States remained the world's industrial powerhouse and everyone wanted to buy its goods. But once Japan and Europe had recovered and the USA began to import more than it exported, increasing sums in dollars were being held by foreign banks – dollars that could be exchanged for gold at any time.

By the mid-1960s, more US dollars were in circulation worldwide than could be supported by the gold in the nation's vaults. Treasuries around the world began to suspect this, but were they to start demanding gold in return for their dollar reserves, the American economy – and with it, the world's – would collapse. The writer and journalist Matthew Hart, in his book *Gold: The Race for the World's Most Seductive Metal* (2013), describes a point at which the USA owed sixty billion dollars when its stocks of gold amounted to just one-fifth of that. Hart quotes one French economist as saying that demanding gold from America would now be a waste of time. 'It's like telling a bald man to comb his hair,' said the economist. 'It isn't there.'

By the time Richard Nixon was elected in 1970, the US economy was on its knees. Americans were increasingly spending their money on Japanese and German cars, America's industry was being hard-hit by international oil price hikes, the cost of the Vietnam War was haemorrhaging cash and inflation was on the increase. Dollars were piling up in the world's central banks while American gold reserves were dwindling. Foreign banks that held vast amounts of dollars began to break ranks and demand gold in return.

Nixon and his advisers saw this as a way of blaming foreigners for the woes of the United States. When, on 9 August 1971, the Bank of England asked for the redemption of three billion dollars of cash into gold, Nixon called an emergency meeting at Camp David. The result, after several days of the world holding its breath, was that Nixon went on TV and blamed 'international currency speculators' for 'waging all-out war on the American dollar.' To defend it, he had ordered the suspension of the convertibility of the dollar into gold.

Everyone holding dollars took a haircut; the dollar's relationship with gold was severed for good, and eventually gold became free to find its own price on the open market.

The people of Yap now use the US dollar for everyday transactions, but for really big or symbolic purchases, the *rai* remains the currency of choice.

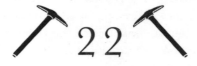

Hangtown

After giving up on Sonora, I followed John Borthwick to Placerville, about eighty miles to the north and just ten from Coloma, where I had begun my journey in the shadow of James Marshall's statue. That seemed like a very long time ago.

I had forsaken my tent and taken a motel room so that I might clean up; I was daggy and malodorous and my beard was too long and grey for my liking. I stood in the afternoon half-light of cheap drawn blinds, looked in the mirror and did what I could to justify the almost two months I had spent looking for gold. This involved convincing myself that I was on a *quest*, and that the supposedly grave and onerous responsibility cradled in that word far outweighed the pouches of nuggets, flakes and dust that I had failed to find. Truly, it did.

Not convinced, I cleaned up and went out to explore.

Placerville is a fascinating little town, not least because that was not its original name; at first it was simply a camp called 'Dry Diggin's' due to the lack of water that covered its ancient – and rich – streambeds, but quickly it became known as 'Hangtown' after vigilantes carried out the first lynchings of the

Gold Rush there in 1849. This was the 'Hang-Town' to which Sarah Royce had referred. In the first year of the Gold Rush, crime was incredibly rare; Hangtown's method of dealing with what little there was soon caught on across California.

Just exactly which criminal act began Hangtown's spate of lynchings is unclear, but two incidents, one in the autumn of 1849 and another in January 1850, are most often cited. In the first, five men were accused of trying to steal a miner's gold dust (or was it, as some accounts suggest, a gambler's winnings?) and were given forty lashes. At the conclusion of the punishment a man in the assembled crowd that had come to watch claimed that three of the men were wanted for a murder on the Stanislaus River. With no more evidence than that, a thirty-minute trial was convened there was a unanimous 'guilty' verdict and the three were hanged from a large oak tree that came to be known, of course, as the Hangin' Tree.

Details of the second incident are also a little sketchy, but most accounts seem to involve a game of cards and some allegations of cheating. One account has a famous card sharp called Dick Crone trying to cut out the heart of a man who accused him of chicanery. Thousands were said to have turned up for his hanging.

In the short term at least, this extreme form of justice worked. Richard Oglesby, an Argonaut who later went on to become governor of Illinois, reflected on the early days of Californian morality and justice thus: 'There was very little law, but a large amount of good order; no churches, but a great deal of religion; no politics, but a large number of politicians; no offices, and strange to say of my countrymen, no office seekers. Crime was rare, for punishment was certain.'

In 1853, Hangtown acquired the first telegraph in the west and its citizens began to feel they deserved to live in a place that sounded less brutal. In an early example of rebranding, they changed the name to Placerville in 1854. But if you walk along its main street today, where there are still fine examples of gold-era architecture, you will find plenty of mentions of Hangtown; the name is spectacularly good for tourism.

Borthwick's first impressions of Placerville were not favourable. The diggings had been exceedingly rich in the early days, when some miners unearthed fabulous finds with nothing more than a bowie knife. Since then, the easy gold had dried up and there was evidence everywhere of the harder work required to find it.

'The beds of the numerous ravines which wrinkle the faces of the hills, the bed of the creek and the little flats alongside it were a confused mass of heaps of dirt and piles of stones lying around the innumerable holes, about six feet square and five or six feet deep, from which they had been thrown out,' Borthwick wrote. 'The original course of the creek was completely obliterated, its waters being distributed into numberless little ditches, and from them conducted into the "long toms" of the miners through canvass hoses looking like immensely long slimy sea-serpents.'

The hillsides had been stripped of all their pine trees to make log cabins, which were dotted about the town wherever men believed there might be gold. There were few considerations for the environment; gold was all.

'There was a continual noise and clatter,' wrote Borthwick, 'as mud, dirt, stones and water were thrown about in all directions, and the men, dressed in ragged clothes and big boots, wielding picks and shovels and rolling big rocks about, were all working as if for their lives, going into it with a will and a degree of energy not usually seen among labouring men. It was altogether a scene which conveyed the idea of hard work in the fullest sense of the words and in comparison with which a gang of railway navvies would have seemed to be merely a party of gentlemen amateurs playing at working *pour passer le temps*.'

So desperate were Argonauts to find plots of land that had not been claimed that they turned to digging up the floor of their own cabins. While Borthwick was prospecting, he used to go to town to resupply and would stay at the home of a doctor friend, but on one occasion he found the settlement where the doctor lived in total disarray. 'The ground on which some of the houses were built had turned out exceedingly rich, and thinking that he might be as lucky as his neighbours, the doctor had got a party of six miners to work the inside of his cabin on half shares,' recalled Borthwick. 'In his cabin were two large holes, six feet square and seven feet deep; in each of these were three miners, picking and shovelling, or washing the dirt in rockers with the water pumped out of the holes. They took about a fortnight in this way to work all the floor of the cabin and found it very rich.'

After working one claim near Placerville for about six weeks, Borthwick sold his share in it and threw mining 'to the dogs'. Perhaps this was because he did not need the money so badly – he had received a sizeable inheritance at the age of twenty-one – or perhaps because he simply realized that, with some notable exceptions, gold mining had become a mug's game.

'There were plenty of men who, after two years' hard work, were not a bit better off than when they commenced,' wrote Borthwick, 'having lost in working one claim what they had made in another, and having frittered away their time in prospecting and wandering about the country from one place to another, always imagining that there were better diggings to be found than those they were in at the time.'

I winced. He might have been talking about me.

Borthwick had always enjoyed sketching and suddenly found himself besieged by miners who wanted their small part in history to be recorded – for they all knew they were in the middle of events of some considerable peculiarity, if not always of significance. 'Every man wanted a sketch of his claim or his cabin or some spot with which he identified himself,' mused Borthwick. 'And as they all offered to pay very handsomely, I was satisfied that I could make paper and pencil much more profitable tools to work with than pick and shovel . . . I invariably found that men of a lower class wanted to be shown in the ordinary costume of the nineteenth century – that is to say, in a coat, waistcoat, white shirt and neck cloth; while gentlemen miners were anxious to appear in character in the most ragged style of California dress.'

I will always remember Placerville as the place where I experienced naked gold fever for the first time. I met there a local businessman who offered to take me prospecting on his own private creek; I shall call him Charlie. He was boyishly enthusiastic and he was kind, introducing me to his lovely family before walking me down to the creek that ran through his property. He already had some high-banking equipment set up and I thought it wonderful not to have to worry about accidental claim-jumping.

We set to work and laboured hard for two or three hours, processing hundreds of pounds of soil down to one-third of a bucket of concentrates. Later we moved several miles south to a beautiful spot on the Cosumnes River, where Charlie put on a face mask and hunted for pieces of gold in the crevices of underwater rocks. I dug in sand in front of a low overhang of rock. I was digging underwater, which is terribly inefficient, and shouldn't have found gold there, but I did – small amounts – and we commented on how there was just no telling where colour might turn up. Charlie found some too, and we put it all in a small vial.

At the end of the day, Charlie handed me the bucket of concentrates and told me I could keep whatever I panned. He was being hospitable and generous, but I declined, explaining that it was my policy – instilled in me by Dave Mack – never to keep more than my fair share. So it was agreed that I would pan out the pay dirt at my leisure and we would meet to share the proceeds the next day.

There were dozens of flakes. It wasn't a bad amount for a day's work, but back in 1849 you would have found more gold on the soles of a miner's boots. The next morning, I took it all to Charlie so we could split it. He took one look at it and his eyes grew dark.

'Not too bad,' I smiled, wondering what was wrong. We were at his office so perhaps he was having a bad day. His staff members were watching. He thrust the vial back into my hand.

'You keep it,' he said.

'I won't do that. Let's split it in half. It was such a good day – gold from two places and a lot of fun.'

'Yeah,' he said, an exaggerated smile spreading across his face. He leant into mine. 'It was great, huh? We did real well – those big pieces you pulled out of the Cosumnes. Didn't expect them, huh? Real big.'

I was momentarily confused. There had indeed been some memorable flakes but we had passed comment on them at the time because they had come from somewhere unlikely – moving sand underneath flowing water way above bedrock – rather than because they'd had any significant dimensions. We hadn't found anything 'big' at all.

Then it hit me. He was making the point that we had found some big pieces of gold so *where were they?* – because they weren't in the vial. He was accusing me of stealing some of the gold we had worked for together. My mind was awhirl. Why would I steal a few flakes of gold when only the night before he had offered to give it all to me and I had refused to take it? This did not make sense, but the smile on his face had grown wider and more knowing. He wasn't a nice guy any more; he had gold fever, and his symptoms were way out of proportion to the amount of gold he must have been thinking of. All the gold we had found that day could not have been worth more than fifty or sixty dollars.

'Here,' I said, 'you take it all.' But he had turned his back on me. His employees were watching, not sure what was happening, and I thought it would be unfair to make a scene in front of them.

Instead, I left and wrote Charlie a letter that night, after a day stewing miserably on the awfulness of the situation. I had lost a putative friendship over . . . what? The next chance I had to visit a post office, I mailed the vial and its entire contents back to him, recorded delivery. I did not want it sullying the rest of my gold.

Aside from the 49ers who made wondrous finds in the earliest days of the Gold Rush, it would be fair to say that it was the merchants and entrepreneurs who made the most money out of the explosion of human activity that surrounded it. You could make a fortune without breaking your back if, like Sam Brannan, you could buy a product for X and sell it for X-plus-Y, with Y amounting to an almost unconscionable mark-up.

Miners needed food, clothing, habitation, equipment, entertainment and transportation, and they had the gold to pay for it. Some of the biggest US household names had their roots in the California Gold Rush.

Before going on to amass one of America's greatest automobile fortunes, John Studebaker used to make wheelbarrows for the miners in Hangtown. Henry Wells and William Fargo, the founders of American Express, made vast riches from the banking services they set up in San Francisco in 1852. Philip Armour, an ambitious nineteen-year-old from New York, travelled to California with a couple of hundred dollars from his parents and opened a business, also in Hangtown, selling meat to the miners; he went on to become the largest supplier in the United States and grew staggeringly wealthy. Somehow, in spite of the fact that his downtrodden employees endured poor pay and dreadful living and working conditions, Armour is still sometimes referred to as a philanthropist.

One Mr Levi Strauss, born in Bavaria in 1829, travelled to San Francisco with his sister, Fanny, on hearing news of the gold finds there. Levi, Fanny and her husband, David Stern, opened their first clothing store in 1853. Levi soon began to make tough clothing for miners who complained that the nature of their work reduced regular habiliments – particularly trousers – to shreds.

The 'waist overalls' that Levi made were further reinforced in 1872 when Jacob Davis, a tailor from Nevada, wrote to him expounding the virtues of small rivets that he had been placing at the parts of trousers that took the most strain. The men patented the riveting process and the rest is history. (What you may not know as you read this in your 501s is that Levi Strauss used much of his wealth for the quiet betterment of others. When he died, one San Franciscan newspaper said of him, 'The great causes of education and charity have . . . suffered a signal loss in the death of Mr Strauss, whose splendid endowments to the University of California will be an enduring testimonial of his worth as a liberal, public-minded citizen, and whose numberless unostentatious acts of charity, in which neither race nor creed were recognized, exemplified his broad and generous love for, and sympathy with, humanity.')

As merchants, Sarah and Josiah Royce slowly prospered. They moved from mining camp to mining camp eight times in five years as gold strikes were announced across the Mother

Lode, and at each step it seemed that Josiah managed to provide more and more replacements for the items abandoned on their long trek west. Describing how she set up home under canvas during one such move, some twenty miles from Sacramento, Sarah's desire to make a real home could not be disguised.

'Our house was of cloth; but the frame of it was excellent,' she wrote. '[It] was not very large; but I contrived to make-believe quite an imposing establishment. In the first place I covered the floor entirely; partly with matting, partly with dark carpeting. One end I curtained off for a bedroom, and by having a trundle bed for the two older children (I had three now) I managed to make room for hanging up the clothing and for standing trunks. The rest of the house, I divided – more by the arrangement of the furniture than by actual partition – into kitchen, dining room and parlor.'

Imagine, after losing everything and only narrowly escaping with your life and that of your child, being able to write of a recovery of sorts measured in tiny but meaningful acquisitions.

Sarah said that her parlour was her pride: 'There was, against the wall, a small table, covered with a cloth, and holding a knickknack or two, and a few choice books. Above it was a narrow shelf with some other books and some papers. There were two or three plush-covered seats, which Mary and I called "ottomans". Their frames were rough boxes which I had stuffed and covered myself.

'The rocking chair, when not required near the stove for baby, was always set in the parlor beside the table, suggesting leisure and ease: but the pride of all was my Melodeon [a type of accordion]. It was said to be the first one that was ever brought to California. It came round the Horn, had been used for a year or two in a Church in Sacramento; and now was, by unusual good fortune, mine . . . There was little time for music during the day, except on Sundays; but at night when the children were all in bed and the store kept my husband away, I used often to indulge myself in the melodies and harmonies that brought to me the most precious memories of earth, and opened up the visions of heaven; and then those bare rafters and cloth walls became for the time a banquet-hall, and a cathedral.'

In contrast with the likes of Levi Strauss and Sarah Royce who 'mined the miners', many of those who persevered with digging found themselves increasingly impoverished and crushingly disappointed.

In September 1850, the dams and flumes built by William Swain and thousands of others on the Feather River to give them access to the riverbed were washed away by early rains. It would be the next summer before gold-bearing soil could be accessed there again, but William was not prepared to wait. Homesick and disillusioned, he decided to sell the few interests he had and return to Sabrina and Eliza.

He was not alone in his disillusionment. According to a report in the *Pacific News* in October 1850, there were fifty-seven thousand miners working the Feather, Bear, American and Yuba rivers at that time, and the vast majority were barely making enough on which to live. In common with many of them, William decided to take a ship via Panama.

'I have made up my mind that I have got enough of California and am coming home as fast as I can,' he wrote. Sabrina must have been overjoyed. Confiding in his brother George, he added: 'I have seen many hardships, dangers and privations and made nothing by it, ie accumulated no property; but if I arrive at home with my health, I shall ever be glad that I have taken this trip. Absence from my friends has given me a true valuation of them, and also it has taught me to appreciate the comforts and blessings of home.'

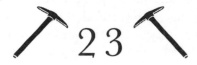

23

Stoned and Confused

I seemed to be moving from place to place more often now as I became increasingly disillusioned with the amount of gold I was finding. After my spat with Charlie I decided to veer from John Borthwick's path, but I found his words ringing in my ears one night as I numbed myself with booze at the edge of a burning campfire on the trail of Sarah Royce.

Borthwick had been in Panama City on his way to the gold-fields at about the same time as William Swain was leaving (Borthwick, remember, did not head for California until 1851). Whilst in Panama, he said that he encountered many would-be Argonauts crossing the isthmus on their way to California and seasoned miners like William returning from their adventures there. Watching these returnees, Borthwick noted that a period spent in California was sufficient to complete a man's education and make him worldly.

'Those coming from New York [heading to the goldfields] . . . seemed to think that each man could do just as he pleased, without regard to the comfort of his neighbours,' he wrote. 'They showed no accommodating spirit but grumbled at everything and were rude and surly in their manners.

'The same men, however, on their return from California, were perfect gentlemen in comparison. They were orderly in their behaviour; though rough, they were not rude, and showed great consideration for others, submitting cheerfully to any personal inconvenience necessary for the common good, and showing by their conduct that they had acquired some notion of their duties to balance the very enlarged idea of their rights which they had formerly entertained.'

Whatever had happened, I wondered, to the 'great consideration' of those travellers past? As the camp fire crackled, three stoned women were firing questions at me while their children bounced around a cavalcade of tents that they had erected, uninvited, around mine. One of the kids was rolling a succession of joints and handing them to my inquisitors.

'You have the vote in England?' one asked.

'Yes.'

'Democracy?'

'Yes.'

'So you have politicians?'

'Yes,' I sighed.

From somewhere in the smooth recesses of my mind, vodka cooling its grey, redundant runnels, the Starship Enterprise hove into view and with it the expression, 'Resistance is Futile'.

'I love England,' said another.

'Me too.' I said.

'Joan of Arc is probably my favourite historical character.'

'Yes,' I capitulated. 'She was a great Englishwoman.'

I had arrived the day before after travelling to the Bear River east of Grass Valley and finding it overrun with RVs, people and dogs. 'Shit,' I thought, 'it must be Friday.'

Past experience had told me to hightail it into the mountains and so I took the opportunity to head farther north and east towards Donner Pass near Truckee, the route chosen by thousands of Argonauts on the last leg of their journey, and the place where the Donner party had resorted to cannibalism. Here the route was so precipitous that wagons had to be lowered down mountainsides on ropes.

For me, at an elevation of five thousand feet on Interstate 80, the going was easy and beautiful; the forested mountains stretched and rolled endlessly under the August sun. But towards the end of 1849, with winter advancing, it would have been astonishingly difficult and dangerous. Many never made it across this final hurdle to the foothills of the Sierra Nevada Mountains. Sarah Royce made it only by the skin of her teeth.

I had tried several campgrounds before settling on one high up near the summit. It was deserted. I chose a pitch under ponderosa pines and positioned my tent so that through its open flap I could see a small stream surrounded by wild flowers. Perfect, I thought, as I set off for a day's prospecting on the North Fork of the American River.

I took lonely tracks deep into the mountains and wound down through primordial forests of oak and pine, quilted underfoot by wild sage and dogwood. After what seemed like an age, I parked my four-by-four and hiked to the river, a sense of anticipation rising in me with the growing sounds of water on rock. The American River here was like a broad stream running deep and fast over huge boulders and enormous aquatic plants. I saw two hummingbirds and stood silently as they hovered from flower to flower. It was some time before I could bring myself to begin prospecting; there was so much beauty and I had to take stock of it.

During the day, I dug in three places. It was always my practice to fill in any holes I made, but here I took particular care to return them as near as I could to the way I had found them. I felt that thousands of eyes were on me but, for the first time, I didn't mind. I found not a single flake of gold.

When I returned, exhausted, to my campsite, a child's tent had been erected in front of my view of the stream. Fifty feet to my right, there were two large tents. A woman ducked out of one.

'Hi there!' she yelled. 'You don't mind my little girl pitching her tent there do you?'

What could I say?

I shook my head and smiled weakly. As the evening progressed, two more women and a dozen children arrived. They put up two more tents next to mine on the pitch that I was paying for.

'Coo-ey!' shouted another of the women, lighting one of two fires that would crackle late into the night. 'You come and have a drink with us later, y'hear.'

It was neighbourliness of sorts but I felt that my solitude had been shattered. Filthy, I changed my T-shirt and drove to Grass Valley, where Sarah and Josiah Royce had finally settled after their five-year existence as Gold Rush nomads. Today, Grass Valley is gentrified and rather chic and Sarah is held in great esteem there. When they had arrived, Josiah was able to build her the home she had always craved. She taught the miners' children while Josiah planted an orchard and sold fruit.

I went in search of the spot where the Royces had settled, hoping to find a house there. It was located at what is today 207 Mill Street, just off the town's main street, but the home Sarah made is gone. Instead, Grass Valley's municipal library, named the Royce Branch, is there. It was named not after Sarah, but her son, Josiah Jr, who was born in 1855. (The Royces also had two more girls while in Grass Valley.) Josiah Jr became a great scholar, studying philosophy at Johns Hopkins University in Baltimore in 1878 before going on to teach at Harvard. He is regarded as one of the foremost philosophers of his day.

In Sarah's account, she doesn't mention Grass Valley by name, describing it only as 'one of the largest and pleasantest of the mining towns quite high up in the Serra Nevada Mountains; higher than Weaverville, our first California dwelling place.

'Here we found ourselves at once in contact with a number of very good Christian people. There were three Churches, all very well attended, and each sustaining a Sunday school. There was also a good sized Public School, as well as one or two social and beneficent societies. Here I seemed to have found, in one sense at least, a rest.'

I felt pleased for Sarah, for her to have found peace and happiness after years of hardship and uncertainty.

Immediately next to Grass Valley – in fact, barely apart from it – is Nevada City and I decided to head there for a drink rather than go back to my tent and the noisy families that now surrounded it. This was one of the first towns that John Borthwick visited after arriving in California and he seemed shocked by the environmental damage that had been heaped upon it by the mining industry.

'It is beautifully situated on the hills bordering a small creek, and has once been surrounded by a forest of magnificent pine trees which, however, had been made to become useful instead of ornamental and nothing remained to show that they had existed but the numbers of stumps all over the hillsides,' he wrote. 'The bed of the creek, which had once flowed past the town, was now choked up with heaps of "trailings" [I believe he meant 'tailings'] – the washed dirt from which the gold has been extracted – the white colour of the dirt rendering it still more unsightly.

'The town itself – or, I should say, the "City", for from the moment of its birth it has been called Nevada City – is, like all mining towns, a mixture of staring white frame-houses, dingy old canvass booths, and log cabins.'

There is nothing dingy about Nevada City today. It boasts not only wonderfully preserved gold-era buildings, but also cherishes the trees that have now grown back to carpet the magnificent surrounding countryside. It has a fine selection of restaurants and bars and I hit most of them. Through a curtain of mist I believe in one saloon I attempted to pay for a drinking companion's whiskey with a pinch of gold.

I slept in the back of my car that night and went prospecting again the next day, finding some small deposits of gold. When I finally went back to the campsite it was dark, but I was spotted before I could duck into my tent and found myself cajoled into having a drink with my neighbours. All three of the women, who probably ranged in age from forty to fifty, were as high as kites.

'So what star sign are you?' asked the youngest, continuing my interrogation. I told her I was a Leo, whereupon a succession of books and astrological charts appeared and it was decided, in between puffs of marijuana, that the alignment of the stars and planets indicated that I was about to enjoy a period of astonishingly good fortune.

'Leos don't believe in astrology,' I joked, pointlessly.

'Oh, neither do I,' said the astrologer-in-chief. Her expression never changed, no matter what she was discussing. Everyone looked at her, and then her friends rounded on her.

'You're crazy about that stuff,' said one.

'Am not.'

'Are too,' said the third.

'Then why d'you haul those charts around with you?'

'Hell, it's just something to do. I'm a sceptic. You gotta show me twenty buckets o' blood or somethin' before I'll believe in anything.'

We sat in silence for a while. I drank more vodka.

'This is my son,' said the astrologer, pointing to the youth rolling the joints. 'He's nineteen. I home-schooled him myself.'

Her son put down a fat spliff and held out his hand for me to shake.

'Can you drive to England?' he asked.

The following week, fed up with my appalling luck (that's astrology for you), I emailed Nathaniel for some advice. How could I improve my haul? His reply suggested I move to the next level and get myself the sluice box that Terry had said – correctly – that I wasn't ready for at the beginning of my trip. He recommended I go to the famous mining town of Auburn, about thirty miles south of Nevada City, and introduce myself to the proprietor of Pioneer Mining Supply, a prospector-extraordinaire called Heather Willis.

I had expected Heather to be a tobacco-chewing version of Margaret Rutherford, but she was nearer to Doris-Day-as-Annie-Oakley on the female spectrum of western movie-ness, not that Margaret Rutherford was ever in a western, or that Heather looked like Doris Day.

Auburn is the seat of Placer County and is the nearest thing you'll find on Highway 49 to a big city. Conveniently positioned also on Interstate 80, the transcontinental highway that links California with New York (loosely following the route taken by many of the Argonauts), it has a population – huge by Gold Country standards – of more than thirteen thousand people. The town's place in gold folklore was assured in May 1848 when a group of Frenchmen camped at a stream in what later became known as Auburn Ravine. One of them, a thirty-seven-year-old named Claude Chana, had been working at

Sutter's Fort and, on hearing of the gold strike at Coloma, rounded up some friends to go prospecting.

When they camped at the ravine, still a days' ride from Coloma. Chana decided to practise his panning skills – and found three nuggets in the first wash. His group forgot all about Coloma after that. During the summer that followed, fifteen hundred miners joined them and discovered, to their never-ending delight, that it was not unusual for a man to pull out gold to the value of a thousand dollars a day.

On one particularly good day, one lucky group found sixteen thousand dollars' worth, the equivalent at the time of forty-three years' pay for one man. More than a hundred years later, a local dentist and sculptor called Kenneth Fox fashioned a forty-five-ton statue of Chana, bending on one knee, as if panning in some river. It is located on the edge of Auburn Old Town, but you will see the image all over Gold Country on leaflets, advertising hoardings, information offices, beer mats, jigsaws, T-shirts and postcards. Miniature versions of it are on sale in just about every gift shop on the Mother Lode.

When I walked into Heather's shop, she was showing two preppy-looking students how to pan. They seemed amusingly terrified by her and bought everything she recommended before quickly shuffling away.

'I don't know why some men can't accept that a woman might find gold mining as exciting as they do, but I guess that's their problem,' she said. I suspected that she had not fully grasped the extent to which some men can be intimidated by an attractive woman.

'Often, guys will come into the store, walk straight past me and go looking for John, my husband, to ask for advice. He's more involved in running the business and manufacturing products, so he'll say, "Go ask Heather," and they'll say, "But I need some information, advice about prospecting," and John will go, "Yeah, ask Heather," and some of them will stand there like they couldn't possibly get help from a female.'

Heather was thirty-eight years old, bespectacled and brimming with enthusiasm. That came across as I mined her knowledge of prospecting, but I imagined she would have been enthusiastic whatever she set her mind to. Her father, Frank Sullivan, was legendary among miners as the man who bought the rights to an automatic panning device and improved it to create a ubiquitous tool called the Blue Bowl Concentrator, which uses centrifugal force in water to separate gold from pay dirt. Frank had been a miner all his life and was delighted when his daughter became as passionate about prospecting as he was.

'He would take me panning and I'd go with him and the Blue Bowl to mining shows,' said Heather. She never stopped moving through the store, rearranging this, fixing that, putting things away; it was a cornucopia of colourful pans, shovels, sluice boxes, pumps, picks, buckets, hoses and generators. I could have stayed there all day. 'Then, when I was eleven or twelve my parents got divorced and I decided I would never go panning again. When you grow up with something like that you take it for granted; you don't appreciate it when it's gone.

'Years later, I found myself helping my dad through his second divorce and I went panning with him again. I was thirty-one and wondered what I'd been missing all those years. When I picked up that pan again and started digging, boy it was hard and there wasn't much gold to show for it, but I felt all those years falling away.'

I asked Heather if she had ever experienced gold fever. Whenever I put this question to a miner, I always left it as vague as that, which meant they could tell me about experiencing symptoms of their own or being on the receiving end of someone else's fevered behaviour. Heather's answer made me shudder.

'I have been in some weird situations and at first they completely freaked me out,' she said. 'Because I run a gold supply store, some customers assume I have some kind of special knowledge about where the gold is, so I've had people following me to see where I'm digging. There have been times when I've been working and had that odd feeling – you know when you sense you're being watched? – and there would be men hiding in the bushes to see where I was digging.

'Those guys are crazy. If there wasn't gold involved would they really think it acceptable to follow a woman, hide in bushes and spy on her?'

I asked Heather which sluice box would be best for a rookie like me and she chose one that cost around a hundred dollars.

'Any suggestions where I should prospect?' I asked. On the scale of mining etiquette, this question put me only one or two notches above the men hiding in the bushes, but Heather didn't seem to mind. She directed me to a gravel bar on the Bear River where I would find a man appropriately named Bear River Gary.

'Kindest guy in the world,' she said. 'He'll show you the ropes. You'll find gold with Gary, for sure.'

Fighting an overwhelming urge to sprint out of the door – and beaming inwardly at the promising twist my quest had taken – I asked how I would recognize this man. I was already convinced he must be *the guy*.

'You won't miss him,' she said. 'He'll be the miner in the deepest hole.'

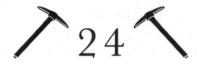

24

Bear River Gary and
the Strange
Usefulness of Gold

I left Interstate 80 where Heather had suggested, at Weimar, and drove slowly past lovely and increasingly isolated houses and homesteads on bucolic lanes with names such as Hope Hill, Quail Valley and Placer Hills Road. I was on my way to meet Gary Shaver, the man – *the guy* – who might finally help me to strike it rich. It was a spectacularly beautiful morning and there wasn't a breath of wind to trouble the pine and madrone trees that lined the banks of the Bear River. I would stew today, but would I stew fruitfully?

During the spring, when snow higher up the mountain ranges melts and roars downhill, the Bear River here is wide, fast and irresistible. Clumsily, I negotiated hundreds of yards of pebbly riverbed that was now dry but which would be impassable during the floods. I was carrying my new sluice box, my pick, shovel, pan, buckets and classifier, and, realizing that this would be the first time in a while that I had not prospected alone, I suddenly felt the need again to give the impression that I knew what I was doing. I couldn't see anyone but I was convinced that hordes of seasoned miners were watching me and shaking their heads.

I followed the sound of the river and came to a gravel bar that was pitted with holes where prospectors had been digging; they hadn't filled them in yesterday because they intended to climb back into them today. These were communal diggings on public land and there were no claims. If someone was in a hole, you left them to it; if a hole was empty, you were free to jump in and begin working it.

It was early in the morning and there were few people around but I still had to figure out which guy was Bear River Gary. I looked over to my left, where some fifty feet away was the biggest hole on the river, about twelve feet wide and thirty feet long and filled with muddy water. In it up to his chest was a whistling man scooping up huge shovels full of dirt and rocks. I went over timidly and established that this was, indeed, the man I was looking for.

'Heather sent me,' I said. He stuck his long-handled shovel into the mud beneath his feet and climbed out of the water.

'Any friend of Heather's,' he said, extending his hand.

Gary was aged forty-nine and once he stood next to me, dripping, I could see he was about five foot ten inches tall, with a wiry frame and a weathered face. He had a grey beard and was short of a few teeth; the ridiculous notion crossed my mind that he was a bit like a young version of an old-timer. I worked with Gary for only a week but in that short space of time he became to me what Walter Huston's Howard was to Humphrey Bogart's Dobbs in *The Treasure of the Sierra Madre*: a guiding hand, a mentor.

'You dig?' he asked, pointing to the hole. I wasn't sure whether he was asking me if I knew how to use a spade or whether I really *liked* the hole, so I nodded, covering both bases.

'Well, okay. Jump in.'

I had been wet on many occasions during my gold-hunting adventures, but this was the first time I was to stand belly-high in brown muddy groundwater; I had concentrated on river banks, not gravel beds. It was also the first time I was to discover just how hard it could be to use a shovel when you couldn't see what you were digging into; in this case, mud and rocks. Behind me, I knew Gary was watching and suppressing laughter but I gave it my best shot. I stuck the blade, unseen, into something hard and slippery, pushed it deeper using my foot on the spade's collar and lifted it up against the weight of the water. By the time it reached the surface everything had fallen off. My shovel was completely empty.

'God damn!' yelled Gary. 'You dig like a Frenchman.'

In the coming days, I was to find that Gary had an international league table of digging proficiency based on perceived wimpishness. By the end of our time together, in a moment that almost brought tears to my eyes – and after I had been disparagingly likened to English, Canadian and Russian diggers – Gary would tell me that I dug like an American. Meantime, and for my first few days with him, my digging was, apparently, French.

We filled our buckets with water and balanced them on whatever flat surfaces we could find at the edges of our hole. The sieve-like classifier would be placed on top of a bucket and one of us would dig while the other shook the classifier, allowing small particles of dirt and gravel and gold to pass through while the larger rocks were thrown away. It was back-breaking work.

On the afternoon of day one, when we had four big, full buckets, Gary led me into the river with my sluice box and set it up in shallow, fast-flowing water. The sluice would allow me to process dirt in much larger quantities than using a pan alone. You simply took a trowel full of concentrates, dropped them on to the sluice at one end, and the heavy sand and gold sank to the bottom, caught in the sluice's riffles and carpets, while the rest was washed away by the river's flow. At the top end was a black ridged rubber mat and immediately pieces of gold began to be trapped there, shining in the sun. Whereas one trowel full of dirt in a pan would take five minutes to wash out, the sluice could sort it all out in a matter of seconds.

As the afternoon progressed, gold flakes continued to appear on the rubber mat, but there would be more in the early evening when the sluice was dismantled and the matting that ran its length cleaned and emptied into a bucket ready for a single panning that would determine whether the day had been good or bad. This was a time of excitement and anticipation and I never tired of it.

We worked hard in hundred-degree heat but were kept cool by manzanita trees above us when we dug, and the crystal waters of the Bear River, wide and shallow here, when we used our sluice boxes. Periodically, we would stop and chat to other miners as they came by to shoot the breeze or seek advice from Gary. I came to the conclusion that I liked the communal aspect of this type of mining; after all, the 49ers worked cheek by jowl during the Gold Rush. Each day, strangers would stop and nod and ask how I was doing and I would say, 'Can't complain,' no matter how much gold was on the rubber mat in my sluice box.

During my first day with Gary I saw so much colour in my pan that I began to regret not finding him at the beginning of my journey. I had travelled two thousand miles, hopping from river to river, but was now no more than thirty from Coloma, where I had started out. I felt annoyed, imagining all the gold I had missed by not coming here sooner. Suddenly, the experience, the beauty and the endeavour of my journey . . . they all meant less, and that left me feeling diminished. Yet the difference in solid gold might have amounted to nothing at all; it was a lottery.

Would I really have exchanged my adventures, the hopes and disappointments, the miners I had encountered along the way, the Siskiyou Wilderness, the Rogue, Merced, Stanislaus, Salmon, Mariposa, Yuba and American rivers, and all the creeks and streams in between, for the chance of more gold here on the Bear River?

I watched my pan glitter in the early evening sun and wondered what sort of person that would make me.

I was being a little unfair earlier when I said that gold was useless; it was practically useless to primitive societies, but modern science has found some highly innovative applications for it. We have come a long way, for example, from thinking that gold was the skin of the sun god Ra to painting it, in layers just 0.000002 inches thick, on to the visors of astronauts' helmets. At this thickness, gold is transparent, yet still effective in reflecting harmful infra-red rays from the sun.

As an electrical conductor, only copper and silver are more efficient, yet gold is the metal of choice in virtually every piece of electrical equipment in your house. The connectors, relay contacts, switches, microprocessor mountings and cable sockets in your phone, laptop, television and hi-fi systems will be golden to some greater or lesser extent. Why? Because a minuscule amount of corrosion in a system that utilizes low voltage and currents can result in disproportionately reduced performance. And, as we know, gold does not corrode.

According to a 2009 United Nations (UN) report on electronic waste, the mining of one ton of gold results in seventeen thousand tons of carbon emissions. The world's present demand for gold in electronics is around three hundred tons a year, resulting in a carbon footprint of just over five million tons. The report emphasized the need to recycle this gold in order to keep carbon emissions down, but its authors hardly needed to spell this out – no one throws gold away forever; there is always someone else just waiting to pick it up. You might discard your phone, but the gold in it – about 0.0012 ounces, worth about fifty cents – will be harvested by any number of reclamation companies. Each year, five hundred million dollars' worth of gold goes into new phones, and in the not too distant future, those phones will all be thrown away.

Following the publication of the UN report, some commentators claimed that there was more gold in a ton of old phones than in a ton of gold ore, assuming the ore had a fairly standard viable return of between one and two grams of gold per ton. One recycling company, Umicore, claimed it could recover three hundred grams of gold from a ton of old phones. Subsequently, the BBC calculated that if all the phones in the world – either in use or lying forgotten in humanity's sock drawer – were recycled, the recovered gold would amount only to the quantity that the mining industry unearthed every twenty-three days. Then there would be no phones left with which to call the BBC to ask how come its journalists have so much time on their hands.

Scientists conducting research into materials at the 'nanoscale', where a nanometer (from the Greek word *nano*, meaning 'dwarf') is one-billionth of a metre (0.000000039 inches), have found that gold behaves differently at the level of the unimaginably small; it becomes red to purple in colour depending on its size (which begs the question: at that scale, would we talk of 'the red standard'; would something be crafted from 'solid purple'?).

Gold has been used in dentistry for more than a thousand

years, because the body doesn't react to it and, unlike teeth, it doesn't decay. This also makes it useful in conditions where the body's auto-immune system goes into overdrive and attacks itself. For example, injections of a gold solution into arthritic joints can dilute this effect and help dampen down inflammation. And, because it is inert and resistant to bacterial infection, it is the metal of choice in a wide variety of medical implants, from pacemakers to weights placed in the eyelids of patients suffering from lagophthalmos, a condition that prevents the upper lid from closing.

Medical researchers believe that gold has a great future helping to fight cancer, both as an imaging diagnostic tool – because of the way it interacts with light in producing image contrasts – and as a 'Trojan horse' delivery system, where radioactive gold 'nanospheres' measuring from two to one hundred nanometres in thickness are delivered directly into tumours. To illustrate just how small this is, a human hair is about eighty thousand nanometres in diameter and a red blood cell about seven thousand.

As an investment vehicle, gold's usefulness is moot; its volatility makes it unreliable. However, as a hedge against disaster, it is still regarded by most financial analysts as the ultimate safe haven.

Once President Gerald Ford lifted the American ban on gold ownership in 1974 and the fixed price for it was abolished (in 1978), levels of demand and extraction soared. Within thirty years the amount coming to market had doubled, from forty million to eighty million ounces per year. As an investment, though, ordinary folk found it impractical; if you wanted to put your money into gold, you had literally to buy it and find somewhere to store it.

In 2004, the introduction of stock market 'exchange-traded funds' in bullion made the whole process much simpler. You could buy gold held by a fund and have it allocated to you in a vault; and you could trade it as easily as any other commodity. Whether this meant that gold was suddenly and usefully democratized, giving the common man and woman access to their own piece of it, or whether it meant gold could now be as easily manipulated as other commodities, is a matter of opinion.

What is certain is that no one knows what the future holds for gold. At the time of writing, demand for it in the West has plateaued, while the Chinese and Russians are buying heaps of it and returning it to the ground in vaults from which it may not emerge for a very long time. According to the World Gold Council, whose job it is to talk up the precious metal, demand in China makes gold a good investment – it forecasts a mind-boggling nineteen per cent increase in demand by 2017 as China's middle class grows by some sixty per cent; at five hundred million, that increase is more than the population of the USA.

The 2014 Thomson Reuters GFMS (formerly Gold Fields Mineral Services) annual gold survey – the gold standard, you might say, of annual gold surveys – registered record demand for almost five thousand tons of gold in 2013, compared with a supply of gold coming from mine production (three thousand tons) and the recycling of scrap gold (twelve hundred tons). That left a shortfall of more than seven hundred tons.

And where demand outstrips supply, prices go up. That's basic economics, right?

Well, not with gold. Gold doesn't go by the normal rules. In fact, in the twelve months before the Thomson report that highlighted that shortfall in supply, the price of gold had fallen by twenty-nine per cent to a four-year low.

In spite of this uncertainty, some investment advisers were arguing that gold was a good bet again. Not because it was worth having as a stand-alone investment but as a hedge against disaster. When the world economy hits a bump in the road and stock markets go into freefall, the gold in your portfolio is likely to increase in value. And, in 2014, with the Middle East falling apart, continued concern over Russian expansionism and the ebola virus running rampant through West Africa, the world appeared to be going to hell in a handcart.

Proponents of gold as an investment will point to its volatility as being indicative of the money you can make on it – and will remind you that between 2001 and 2011 it made gains of almost 450 per cent. What they rarely tell you is that, when adjusted for inflation, it is worth less than it was in 1980. If it had kept pace, it would be worth a thousand dollars an ounce more.

In April 2014, Robin Ash, a *Times* reporter based in London, asked five city analysts whether gold was a quintessential safe-haven bet. Four said yes; one said no. The one who said no predicted that the price of gold would fall by a further twenty per cent on top of the big losses it had made during the previous year, because it simply could no longer be relied upon.

The four 'yes' votes came with miserable qualifications. Russian military action in Ukraine, the potential for future inflation in the euro zone and market currency turmoil could all be good for gold – yet they were all factors that spelled bad news for the world economy and the ordinary man and woman in the street.

And that's the thing about gold. To really believe in it, you have to be a God-awful pessimist.

25

The Curative Powers
of Colour

Bear River Gary seemed to find something to be happy
about during every moment I was fortunate enough to
have him as my partner. He was strong like a terrier and would
dig deeper than me, work longer and still find time to laugh
and joke and talk endlessly, and I think this was because he
half-expected to drop dead at any moment.

In June 2004 Gary had suffered a heart attack. It nearly
killed him, and then he had three more. Since then, doctors
had fitted him with a miniature defibrillator inside his chest to
kick-start his heart if he had another. He peeled back his shirt
to show me the scar.

'Could go any minute,' he said, cheerfully. 'At first, my
ex-wife, Lori, and my stepdaughter, Crystal, weren't too happy
about me going into the wilderness to prospect, in case I had
another heart attack. That was just eighteen months ago and I
was unfit, depressed and way overweight. But then they saw
the benefits and realized how much good this was doing me. I

move fifty-five hundred pounds of dirt in a typical five-day week.

'My doctor can't believe the change in me. He used to tell me to be careful of those hundred-pound buckets but now he just says, "I don't know what you're doing Gary, but you just keep on doing it." I'm fit as a fiddle and I couldn't be happier; come down here each day and find pieces of gold. If I'm going to kick the bucket, then what a way to go, in a beautiful river surrounded by trees and mountains. Better than dying behind a desk or in front of the TV.'

Although he enjoyed modest success, I came to the conclusion that Bear River Gary was not *the guy*. He had become something of a prospecting authority in less than two years, but his finds were not the stuff of legend. Instead, mining had bestowed upon him something else. He told me that after his heart attacks he had been withdrawn, overweight and desperately sad, but since becoming a gold miner he was a changed man, so full of energy, so consistently upbeat. Subsequently, I was to meet other miners who told me about the curative powers of gold, or at least of looking for it. Aside from Gary, I met several heart attack survivors, two people recovering from cancer and many more who had been lost in depression. They all told me that prospecting had saved their lives.

None of these was more inspirational than Mike Morgan here on the Bear River. The first time I saw Mike, his head and shoulders were above water in a hole and I assumed he was standing below trying to dig down to bedrock. He saw me and waved. It was early and the sun had not risen above the trees, so the water in the holes dotted about the gravel bar would have been extremely cold. Crazy guy must be freezing, I thought. Or drowning.

Mike was aged thirty-eight and handsome, with a broad white smile that never seemed to leave his face. He was married to Joanna, who was originally from Birmingham in the

UK, and they had two sons, Greg, who was ten, and James, who was five. They had moved from Sacramento to Grass Valley the previous year to get closer to nature; it was a decision that was to change Mike's life.

He had been a construction worker and a keen sportsman when, ten years earlier, he and a colleague were checking the pressure in a fire hydrant and the test gauge blew off, forcing what was later calculated as seventy-one thousand pounds of water travelling at great speed to hit Mike in the small of the back, catapulting him over a distance of thirty feet in the process. His spinal cord was, in his words, 'kinked and folded over', resulting in severe damage; four vertebrae were fused together, he had titanium rods inserted into his body for support, and at first he was completely paralysed from the waist down. After ten years of therapy, he had regained some movement in his right leg, but hardly any in his left.

As I got closer to Mike, I could see that his hole was not deep at all. It was maybe three feet, but he was sitting in it and digging. To get here, he had parked his modified vehicle about a quarter of a mile away, steadied himself on crutches whilst balancing his sluice box, buckets, classifier, spade, pan, pick and trowel on a cradle across his back, and negotiated the dry riverbed, its rocks and shifting cobbles, before unpacking everything and lowering himself into this freezing water.

'How's it goin'?' he yelled, holding out his hand. I asked if I might sit down with my feet in his diggings. 'Sure,' he said. 'Make yourself at home.'

After coming round from surgery, Mike had been able only to wiggle a single toe on his right foot and, for a man with a seven handicap at golf, that had been hard to handle.

'I'd always said that if I ever got paralysed, just put a bullet in me,' he said. 'But Joanna refused to let me feel sorry for myself and she'd kick my ass if ever I began to allow myself to start giving up. It was tough for her – Greg was just one year old and there she was with a baby and a husband who couldn't walk and who was pretty depressed.

'But after a while, I began to realize that there was more to life than walking. Hey, I had my wife and my child, then another son. I had an enormous amount of support, and one day I thought, "I'm going to make this work. I'm not going to let my disability define who I am and what I'm going to do." And Joanna was amazing; she stuck by me one hundred per cent from day one, no questions asked, being strong, making me stronger. Boy, did I find a keeper.

'I'm not saying it's always been easy. There are times when I'm rolling around playing with my boys and I see other dads running and walking with their kids. Of course that hurts, but I try to do everything with my boys that they do with theirs – I'm just a bit slower, that's all.'

Shortly after the family moved to Grass Valley in a house from which Joanna runs a bakery and wedding cake business – ('Man, our house smells of cakes every day!') – Mike took the boys down to the Bear River for a swim and saw Gary and the other miners at work.

'Somebody showed me some gold they had found and I said to myself, "I could do that." I had some old pans lying round the house so I got myself down here, found a hole, climbed in it and started panning. Then the guys showed me how much more I could process with a sluice box and I was hooked. I come down here at least a couple of times a week, sometimes four.

'I've been doing it for about four months now. At first, Joanna thought I was crazy, but I've lost six pounds, my body is toned and I feel positive. I think my balance has improved, too. I love being out here in the wilderness and I meet new people every day. And there's the gold; I've found half an ounce so far and I regularly pick up one and a half grams a day

– it won't exactly make me rich but, hey, that's about sixty dollars!'

The first time I met Mike I had risen early because I couldn't sleep; I was missing home. I looked at him and imagined his boys and Joanna, their house smelling of hot sponge pudding and cinnamon. This was what made him truly wealthy; he didn't need his gold.

Thoughts of calling it a day came more regularly now. I was finding more colour but, like Mike's haul, it was only a gram here, a gram there. Spread thinly, it looked good in the pan but it wasn't enough to change even the smallest aspect of my life. From the moment I had pulled out my first colourful, almost weightless flake of gold from the Klamath, I had secretly known this. But that didn't make me feel any less of a fool for finally admitting it.

A week after I met Mike, I found myself twenty miles away, standing in a steel cube frame measuring seven feet by seven feet by seven feet. Raising my hand above my head, I would have been able to touch the cube's ceiling had it had one. It was two skips or three hops wide, and – most comforting of all – it was delightfully unimpressive.

The cube frame was located in the courtyard of the Empire Mine at Ophir Hill on the south-eastern outskirts of Grass Valley, close to the place where Sarah Royce had settled with her family. Now a museum and 850 acres of gardens, Empire Mine operated between 1850 and 1956 and was the richest lode mine in Californian history, producing almost six million ounces of gold, worth about eight billion dollars today, before it became unprofitable and was closed down.

During that period, tens of thousands of men – including large numbers poached from England's most prolific tin mines in Cornwall – had dug 367 miles of tunnels, some to a depth of eleven thousand feet. The mine operated three shifts a day, seven days a week, 365 days a year – except for three days once given over to celebrate the wedding of the owner's daughter – and it utilized the latest equipment available at the time. Geologists and mining engineers supervised the blasting of new tunnels to follow the tendrils of the Mother Lode and no trick was missed, no opportunity passed over, no lesson wasted, no skill ignored, no scintilla of a whiff of an edge not exploited in the incessant pursuit of gold.

Yet all of the precious metal mined by those thousands of men with their costly equipment and snarling corporate voraciousness – over a period of 106 years – would fit into this frame measuring seven feet by seven feet by seven feet. All of it.

I looked at the flakes of gold in my vial, mined by an inexperienced city slicker with two left thumbs in less than two months, and I no longer felt so bad.

When this feeling of faux smugness passed, it was replaced by another, of unbridled astonishment. Until now, I felt I had understood something of the power of gold and its value, yet I had not come even close until I stood inside this seven-foot cube; I imagine that is why the museum had had it made, to put the preciousness of this metal into some kind of perspective. Something this compact had fuelled 106 years of economic activity – had provided a living for thousands of families; had returned dividends for investors; covered the cost of plant, machinery, research and technology; *and turned a profit*. It made the Bourn family, who owned the mine for half of the twentieth century, fabulously wealthy.

The profitability of Empire Mine rose and fell, but an

indication of how rich it was can be gleaned from 1867 records that put the cost per ton of fetching ore from the ground at eight dollars and sixty cents per ton, and the gold yield from each of those tons at fifty-nine dollars. Historians say that the noise from the forty-stamp mill that separated the gold from its quartz shell was notoriously deafening and could be heard more than three miles away. When it stopped after a strike in 1956, they say the silence drove some people crazy.

I was shown around the museum and taken into one of the old mineshafts by Jeff Herman, a former supervising ranger at the facility. Jeff's enthusiasm for the mine was infectious. He let me sit in one of the trolleys that propelled the miners down to their workplace at a speed of six hundred feet per minute and I felt vicariously proud that my English countrymen had probably dug the shaft. Cornishmen were valued for their experience of mining for tin underground, and they were recruited for more than three generations. If you sat where I was now, in 1860, you would most probably have heard thick West Country chatter floating on the air, issued from mouths half-filled with Cornish pasties.

After a time, Jeff led me towards a bench on which a couple were sitting, and introduced them as Gene and Cathy Meyers. Gene was aged sixty-four and Cathy was a year older. In his thoughtfulness, Jeff had asked them to come and meet the English guy with a passion for gold. I had not expected to be introduced to anyone and found myself a little tongue-tied. Cathy smiled but Gene looked as puzzled as I was by the introduction.

'If you want to learn about gold prospecting, I think you should talk to Gene,' said Jeff, and then he went about some other business. I smiled at Cathy and mumbled that perhaps it might be nice to walk in the gardens and sit near the Bourn family's country residence, Empire Cottage, a mock-Tudor manor house with a rather chintzy arts-and-crafts interior.

Here, the family had spent summers in the lap of luxury while their employees laboured thousands of feet below.

After a time, Gene and Cathy sat on a bench beneath ponderosa pines and a wide and beautiful cork tree while I positioned myself on some grass at their feet. Cathy did most of the talking, up to and including the part where she explained that Gene was suffering from Alzheimer's disease, and then Gene spoke, telling me about a family picture he had of him holding a gold pan at the age of three and of the times as a boy when he and his father, Bob, would disappear deep into the wilderness to prospect.

Sixty-four was a young age to be suffering from Alzheimer's and Gene had been diagnosed five or six years earlier. Later, in some deep forest, he would tell me that his doctors had given him only a couple of years to live, and I cannot fully describe how sad that was. He was a picture of health, with a full head of silver hair, slim and fit in a blue T-shirt and dark-blue slacks. Cathy, sitting beside him and never letting go of his hand, was dressed in white beneath a floppy cream-coloured hat. She wore spectacles and smiled encouragingly, constantly offering up stories about Gene, prodding him gently to finish them, like a proud and loving *aide memoire*.

Gene's father had been a machinist and he had a friend who was an engine designer. During the 1970s, the engineer had kept coming up with plans for pumps that could be used on the back of small rafts to suck up river sediment while simultaneously providing air to a diver below, and Bob would try them out. These formed the heartbeat of some of the first portable dredges that had only recently been banned in California, and Gene and Bob had been the guinea pigs who went out into the wilderness to try them out.

'I saw the first colour in my pan when I was six or seven, probably somewhere near Oroville,' Gene told me, with perfect recall. 'And I reckon we first started dredging in Spoon

Creek, near Feather Falls, when I was a teenager. We used to haul the dredge to some pretty wild places and we'd camp there, drink from streams and catch fish to eat. We found a lot of gold, but it wasn't its worth that excited me; it was just being out there looking for it.'

Gene said that one theory his doctors had for the early onset of his symptoms was that they could have been caused by those long expeditions into the wild. Rivers were more polluted back then, he said, so eating fish and drinking water from streams and rivers could have exposed him to dangerous levels of mercury. I asked him if that possibility sullied the memories.

No, he replied. He wouldn't change them for the world.

'The biggest nugget I ever found weighed five ounces,' he went on. 'I got that in 1987 in Dogwood Creek, near La Porte, but I had better days. In the summer of 1983, on the South Fork of the Feather River, I was looking for a good place to prospect – we had hauled the dredge down mountainsides into a canyon about a mile deep – and I was floating in the river when I saw something shining on the bottom. I dived down and it was a two-and-a-half-ounce nugget. Then I dived down again and found one weighing three ounces. When I went down a third time, I found another, weighing one-and-a-half. Brought out nine ounces that day and I haven't heard of anyone in modern times finding more.'

Cathy squeezed Gene's hand. She had a photograph of some of the gold he had collected. It was a picture of some scales containing at least fifty sizeable nuggets, including the weighty ones he had pulled out of the Feather. Together, the collection would have been worth a fortune but it was only a fraction of what he and Bob had found over the years. During one season, they pulled out twenty-five ounces; in another, thirty-three.

'He and his dad were legendary,' said Cathy, proudly.

'They found a lot of gold but we used to hear stories with some pretty exaggerated claims about them finding even more. "Old Man Meyers and his son pulled out eighty ounces of gold! . . . Old Man Meyers and his son filled a quart jar with gold out of such-and-such a river . . ." We used to laugh at some of the stories.'

I looked up from the photograph and blinked. I had spent almost two months hearing about *the guy* and had arrived at the conclusion that, outside of the original Gold Rush, he simply did not exist.

Now here, in front of me, I had him, *the guy*. And I wasn't about to let him go.

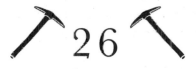

26

The Guy

Gene had two dogs, one called Feather and another called River, and that did not surprise me one bit. Today, Feather, a border collie–chow cross, was to accompany us on a journey to the South Fork of the Feather River to try to find the spot where, almost exactly thirty years ago, Gene had pulled out those nine ounces of gold in a single day.

When I first met the couple, Cathy had expressed regret that Gene's biggest disappointment since falling ill was that he could no longer get into the wild on a regular basis. The couple had two grown-up sons but, like everyone else, they had daily commitments, and Gene would have loved to get up and out there every day. Cathy, once Gene's enthusiastic prospecting partner, could no longer hike and climb to his favourite places because she suffered from knee problems. And, because of Gene's illness, he'd had to surrender his driving licence the previous year. Each day, he would take Feather and River on a circular walk near their home in Chicago Park, a leafy community south-east of Grass Valley, but he found this a good deal less than fulfilling.

'Would you come out with me?' I had asked the previous day. I fully expected Gene to say no but the reaction was startling. His face lit up and the look of mild confusion that he bore with such dignity evaporated, to be replaced with a broad smile. I wondered whether Cathy might have reservations but she was every bit as pleased as he was.

'Sure,' he had said. 'You want to go right now?'

Instantly, I began to wonder how much of my offer amounted to altruism and how much to the possibility that some of Gene's luck might rub off on me. I tried to think of another time when I had offered to take a sixty-four-year-old person with Alzheimer's disease out for the day and I had to admit that there wasn't one.

'You'd do that for Gene?' asked Cathy. 'That would be wonderful!'

I had found myself impressed by Gene and drawn to him – I was enjoying his and Cathy's company and their encyclopaedic knowledge of serious prospecting – but was I doing this in the hope of finding gold for myself? Was I *using* him? The thought horrified me because I really did not know and I found the possibility shameful. The fact that I could not judge one way or the other suggested that my moral compass was depressingly skewed. Gold, I thought, bloody gold.

We had decided I would pick up Gene early in the morning, while overnight I convinced myself that his company, the hike and the education I would receive at the hands of a master prospector were my true rewards for going out into the wilderness with him. Nevertheless, I figured it would do no harm to pack my pick, pan and trowel in my rucksack, just in case.

It was a cold morning and it was to remain cool until noon. Gene appeared a little confused to see me on his doorstep and he spoke to me quietly and with a reserve that I could not quite fathom. When he talked, he seemed slightly unsure of the weight of what he was saying, meaning that statements sometimes sounded like questions, but I had no idea whether this was to do with his condition or whether it was just Gene's way and always had been. If it was the former, then I was concerned that I was adding to the daily effort of his getting by.

Gene asked if I had brought proper hiking gear with me, some food and lots of water and I told him I had. We would be tackling acres of poison oak, so long trousers and sleeves were a must.

Already, I was relieved that the morning temperature had remained bearable. We set off driving north, with Gene giving me clear directions towards the Feather River east of Oroville near Forbestown, a tiny community named after a man who had opened a store there in 1850. It was a ninety-minute journey from Gene's house.

We drove for long spells without speaking but the silence wasn't awkward. Instead, I had the feeling that Gene was concentrating on the route. From time to time, I would ask questions and found that he was good at answering generalities but not specifics.

'I can't describe what it's like,' he told me at the end of one long silence. 'It's as if you know there's something there and you're trying to take a hold of it but it stays just out of reach.'

Gene had been a maths teacher at a school for problem kids and I guessed that he would have been a popular member of staff. He was quietly spoken and patient, valued qualities in such an environment. I could only imagine the impact that Alzheimer's would have on the mind of a mathematician.

As we came closer to Forbestown, his concentration intensified. 'Take that track there, to the left,' he said. 'We'll find a path going off that just up ahead.' But we didn't and a look of perplexity spread across Gene's face. What on earth was I putting him through?

After five minutes of weaving down a quickly vanishing trail, he asked me to turn around. 'This isn't it,' he said. He seemed annoyed with himself. I drove back to the road and continued, Gene scanning intently for the correct turn-off. I knew we wouldn't find it. What was I thinking, pressing him to remember the scene of his greatest triumph – some thirty years ago – when memory was the thing that eluded him most? It was inconsiderate in the extreme; I was being cruel.

I was about to stop and offer to take Gene somewhere else, somewhere random, when suddenly he shouted: 'There! That's it – turn left here.'

The road wound for several minutes before he told me to pull over underneath a collection of tall oaks and ponderosa pines next to the faintest of paths. I glanced at him to see a look of certainty and pride on his face. 'What are you waiting for?' he said.

Feather jumped out and we took our rucksacks from the back of the four-by-four. Gene had brought a stick and he reminded me to take mine, then off we set. 'Watch out for rattlers,' he said.

At first, the terrain was easy, pitching up and down but mostly down, through thick and ancient forest and there were signs that other people had been there, though not recently. Gradually, the way became steeper and all evidence of pathways and people disappeared, to be replaced by a foot or more of fallen leaves and the young ferns that pushed their way through them. As the gradient increased, I had to hold on to young trees for fear of falling, but I lost my footing and plunged through layer upon layer of poison oak. Gene was ebullient. The look of helplessness on his face had fallen away and now he exuded only confidence.

'Look at this,' he said after a time, pointing first at the forest floor, then at the twigs and branches that surrounded us. I couldn't see anything out of the ordinary. 'Pristine,' he said. 'Nobody's been here for years.'

After half an hour of hiking down, we emerged on to a wide

promontory with a view of the opposite side of the canyon we were descending. It was densely covered in fir, madrone, oak and pine punctuated only by craggy outcrops and fallen shelves of granite, some the size of cathedrals. Eagles hovered below us. There were no paths or sections of open ground and I wondered whether, from over there, the side we were traversing had the same appearance. If it did, then we were in truly wild country. I could hear the far-off roar of rushing water but there was no view of the river and there would not be for some time. It was a mile below us and we were scrambling down to it at a gradient of forty degrees and more. I mostly slid across leaves on my backside, using rocks and young trees to slow the speed of my descent; Gene remained upright, nimble as a mountain goat. He seemed to know the way even though there was no trail to follow.

We hiked like that for an hour and more, sometimes hanging on for grim death, sometimes zig-zagging to stagger the slope, while the sound of unseen water grew louder. As the hands on my watch passed noon, the cool of morning was crushed by the oppression of the sun, no longer hidden behind the mountains but above us. Gene stripped down to his T-shirt but I remained covered. 'Poison oak's a bitch,' I kept hearing Craig say, way back in Happy Camp.

Our route was littered with fallen tree trunks, but after my first attempts at using them as way stations to slow me down, I avoided them; each one had completely disintegrated into sawdust, showering me with brown powder, fungal spores and wood lice. Instead, I continued to put my trust in young saplings and the braking power of my behind. Periodically, Gene would stop and put his hands on his hips, looking back uphill to wait for me. Feather would catch up and sit by his side.

'You okay?' he would inquire. He looked fresh and happy and full of energy. I was covered in dry leaves and ferns, thorns, wild blackberry stains, wood lice and rivulets of sweat.

'Tickety-boo,' I said.

I had brought four-and-a-half litres of water and had drunk one-third of it already. Gene hadn't had a sip of his.

I asked him if we might sit for a while. I drank more water and caught my breath while he went down some way to have a scout and then climbed back up to me. Eventually, after an hour and a half, rotting vegetation was replaced by huge granite boulders and black volcanic rocks and these led to a wide shelf below which was a bend in the river. The water fell in stages, travelling calmly to leave huge and inviting pools at intervals of several hundred feet.

'There it is,' shouted Gene, pointing down. 'That's where I found my nine ounces.'

There was not the slightest trace of doubt in his voice.

'I was just looking on the bed of the river, fanning material out of the way with my hand when I saw something and thought – what the heck's that?' he said when I caught him up. 'I picked it up and it was the two-and-a-half-ounce nugget. I just thought, "Wow! Things like this just don't happen." Then I found the three ounce and then the one weighing one-and-a-half ounces. There were other nuggets there too and I collected them and finished with nine ounces of gold in twenty-four hours. It was the most productive day of my life. I had told my dad I thought this would be a good place to prospect, and later he said, "Well son, I guess you were right."'

After a moment's rest, we moved upstream and climbed over boulders that skirted some of the loveliest stretches of water that I had seen anywhere in the world. Elephant-eared leaves of Indian rhubarb hung over pools as clear as flawless diamonds. The sense that we were alone – and in a place unseen by anyone else, possibly for years – added layer upon layer to the mystery and beauty of it.

William Swain had prospected on this fork of the Feather in 1850. It could not have been far away, and the thought made me shiver. I imagined Gene and William sitting around a camp fire, swapping tales of gold, and I felt they would have made fine companions.

Explaining why he chose to prospect here, William wrote to Sabrina in January 1850: 'From all that [we] had learned and that we could hear, we judged the South Fork of the Feather River to be the most likely to yield a pile another summer, for the following reasons: the main part of the Feather River and all the southern rivers have been overrun and consequently the best and richest placers found and worked.

'The South Fork of the Feather River was reported to be rich and the gold on it coarse and not much worked. There is good timber for building (not the case on many of the streams of California), which with us is an important consideration as we believed our health next summer depended upon having dry, warm and comfortable habitation during the rainy season.'

William and his friends bought provisions at Long's Trading Post, located in today's Butte County, and hiked twenty-five miles east over mountainous terrain similar to that just traversed by Gene and me. They built their cabin and, the following month, William was upbeat, writing to Sabrina: 'You will of course ask: have you done nothing yet in mining? Yes, I have done considerable. I have panned along the banks of the river with various success. My first day's work in the business was an ounce; the second was thirty-five dollars [about two ounces], and the third ninety-two. I picked up a lump worth fifty-one, which cost me no more labour than stooping down to take it up. But such days' work as these are not a common thing.'

As we know, William and his friends worked for five months to dam the river only to find that its promise of plentiful gold was never fulfilled. More than a century later, it was to bestow more luck on Gene.

Periodically, bewildered by the elegance of my surroundings, I wondered if I might see a scrap of paper, a discarded bottle top or a disintegrating cigarette butt, anywhere, but there was no waste to speak of save nature's atrophy, which isn't waste at all. At any moment, I expected Gene to announce the spot where we would dig or crevice or enter the water, but he just hopped from rock to rock as the memories came flooding back.

'See that boulder?' he yelled. 'I moved that to get at some gold. Rigged up a grip hoist and cables to those trees and hauled it out of the way. Helluva job. Took me two days. And there! I pulled some nuggets from there but I had to move that fallen tree trunk out of the way first.'

When one side of the river became impassable we hopped across rocks and waded into the water to reach the other. The smell of wild sage hung thickly on the air with sweet, rotting vegetation. Nothing I could do would persuade Gene to eat his lunch or drink some water, but he returned to me time and time again with handfuls of the fattest, sweetest blackberries I had ever tasted. (Cathy told me later that one of the causes of death among late-stage Alzheimer's sufferers is a loss of desire for food and drink.)

Some way upriver we switched banks again and climbed up to a granite shelf that hung over the water. There were blackened traces of fire on the rock and some short distance away someone had dug into the bank below a tree.

'That was me,' said Gene. 'I came back here ten years ago and camped for a while. I tried digging there – you sometimes find gold in the roots of trees – but there was nothing.'

I hadn't realized that he had been back since the day of his big find but it didn't matter; returning successfully to such an isolated location in his condition would have been a remarkable feat even if he had been there only a week earlier. What was clear – and seemingly important to Gene – was that no one else had camped at this spot since. He used to prospect during the day and sit on the ledge, casting his line into the river in the late afternoon for something to cook on his fire in the evening. These were Gene's halcyon days. His face, remembering them, was a picture I shall never forget.

As the afternoon wore on, he showed me more of his favourite and most bountiful places and explained the river craft behind his decisions to work them. I listened but mostly said nothing as Gene came to life in his own personal heaven.

Not once did I ask whether I might take out my pan and do a little prospecting. I saw no point. It was as clear as the nose on Gene's face that today we were mining something much more precious than gold.

The climb out of the canyon made me feel nostalgic for the hike into it. How could I possibly have felt so bad going downhill for a couple of hours? Compared with the four-hour ascent it was nothing.

We stopped regularly – I had to in order to avoid collapsing – and this gave me a chance to talk to Gene about his illness. He said he had been diagnosed 'about six years ago' and the doctor who made the diagnosis gave him approximately eight years.

'Eight years until you lose your memory, your sense of self?' I asked.

'No, eight years until I die.'

I felt flattened. Gene was fitter, seemed healthier, than me. If the doctor was right, he had two years left. 'I'm sorry,' I said.

He was picking up long grass and skinning it with his thumb.

'I can't complain because I've had a wonderful life and I've done everything I've ever wanted to do,' he shrugged. 'So I'm not worrying about it. I can't say I feel happy about it, because I don't, but I still have Cathy and I can still do this. Look around you – what more could a man want?'

At the top of the canyon Gene remained more alert than me; he saved me from standing on a rattlesnake and saw it off safely into the undergrowth. We gave Feather some water and I drove to a gas station, where I bought Gene a sweet drink and a packet of crisps, refusing to take him home until he had drunk and eaten. I figured it was the only way I could get sugar, salt and fluid into him. We chatted for a while but gradually a stillness descended on him and he became quiet again.

Our journey home was spent in silence. Gene seemed to have retreated into himself once more. Cathy had burritos waiting for us and I ate mine hungrily while we talked about our outing, but I could not stop thinking about Gene fading away. There was some consolation at the end of the day when Cathy explained that he was having difficulty with numbers and that he did not have only two years to live; it could be more like ten.

Months later I contacted Cathy to see how Geno, as she called him, was doing. He had deteriorated psychologically but was still in perfect physical condition, a contrast that had led to some frustration, anxiety and depression on his part. He still hankered for the outdoors.

'He's been having a hard time, and that makes it hard for me, his family and friends too,' she said. 'Outwardly, he still looks like Geno but it is not really him in there most days and I miss him.'

I still hark back to the day I spent with Gene – *the guy* – and smile from the sheer joy of it, from watching something vital return to him for a few hours. If I were granted a wish today, it would not be for gold, but for Cathy to have been there.

In the absence of that, all I can say is that her Geno was happy, flitting through time at the speed of light.

In fact, at the speed of memory.

27

The Elephant

If the California Gold Rush told its participants that they could be somebody, or at least be somebody different, then it also gave the United States of America a glimpse of what it could be as a nation. During the Gold Rush, commerce was conducted with a handshake and at breakneck speed; towns burned down and were rebuilt in a matter of days; rules were torn up and rewritten for the better; codes of morality slackened, leading to the concepts of individuality and freedom of expression. When men returned to their homes and farms in the east, the stories they told generated wonderment in their communities and often restlessness too. Many thousands, disillusioned with the gloom and plodding predictability of the old ways of life, returned to California, but this time with their wives and children, not as Argonauts but as farmers and builders, lawyers and physicians, to forge the state proper.

The sudden growth of California alarmed the southern, slave-owning states. As early as 1849, would-be Californian statesmen had come together at a convention in Monterey to draw up a constitution and apply for admission into the Union;

their constitution forbade slavery. Until then, there had been thirty states – fifteen free, fifteen slave. The admission of California would swing the balance of power the way of the free states, and that was something against which the southern politicians fought hard. They proposed splitting California into two halves, one that recognized slavery, one that did not, but the idea never gained traction. California was admitted to the Union on 9 September 1850. The Argonauts celebrated with fireworks and champagne.

With civil war still a decade away, dozens of ships now plied the seas around Cape Horn or to Panama and Nicaragua for those who chose to cross the isthmus. Conditions aboard passenger vessels changed for the better and ticket prices fell with competition. By the time the Gold Rush petered out in 1855, you could get to San Francisco from New York in just three weeks. Overland routes – at least during the spring and summer – became more bearable with the building of bridges and the establishment of hotels, general stores and blacksmiths along the routes. Then, after the war, the transcontinental railroad was opened and California's sense of isolation came to an end. The state boomed and never really stopped. Today, if California were a country, its economy would rank eighth in the world.

As he had feared, the Gold Rush destroyed John Sutter's dreams of a prosperous New Helvetia and eventually it left him in financial ruin. Even more painful than this was the treachery visited upon him by one of his children, August, who had tracked him down from Switzerland to Sutter's Fort in September 1848, just as the Gold Rush was gaining momentum.

Then aged twenty-one, August had been only seven when Sutter abandoned him, yet father and son managed a civilized reunion of sorts, a reunion from which Sutter – once again in debt – sought to gain an advantage. He signed over most of his property to August to avoid it being seized by creditors, but when the boy began to build a city near Sutter's Fort to exploit the Gold Rush and the people it brought, he called it Sacramento and not, as his father had wished, Sutterville. The choice of name was a dagger through John Sutter's heart.

The old man sank into alcoholism and spent almost two decades petitioning Congress to reimburse him for the damage that the Gold Rush had inflicted on his business empire. In 1880 a bill to give him restitution in the sum of fifty thousand dollars was debated. It was expected to pass comfortably but Congress adjourned before a vote on it could be taken. Two days later, on 18 June, Sutter died. He might have found some consolation in the fact that his name is given to Sutter Creek, a small town in Gold Country, and to a central street in downtown San Francisco. Of course, we will never know.

While Sutter's fortunes crashed, Sam Brannan's soared. The two disliked each other intensely, even though they had been occasional business partners, so it must have added further to Sutter's pain to see Brannan prosper. After moving his store from Sutter's Fort to the new Sacramento, Brannan's takings often topped five thousand dollars a day. He had a hand in everything, from politics to policing, warehouses to mines, hotels to wharves and docksides. He established the first overland mail service to the east and shipped the first steam locomotive around Cape Horn in the other direction.

But it wasn't to last. Brannan sank his fortune into the construction of a spa resort built on hot springs in the Napa Valley, but, a hopeless drunk, he gambled much of it away and was forced to liquidate the rest when his wife, Ann Eliza, divorced him. (Brannan was regularly inebriated; at the opening of the resort, he had intended to describe it as the 'Saratoga of California', likening it to the spa town in New York State. Unfortunately, it came out as the 'Calistoga of Sarafornia' and the name stuck. There is a small museum in Calistoga where you can read about Brannan's exploits, smart and crazy alike. E Clampus Vitus has a chapter devoted to him.)

Like his adversary John Sutter, Brannan sought solace in even more drink. He tried to revive his fortunes with further investments in Mexico but they, too, went sour. He married twice more and settled with his third wife, a Mexican, in Escondido, north of San Diego, where he died in May 1889. By the end, she had left him too and there was no money for a funeral. Sixteen months after his death, a nephew had Brannan's embalmed corpse removed from an undertaker's vault and given a decent burial.

Sarah and Josiah Royce never left California, though they did move from Grass Valley to San Francisco after twelve years in order to further their children's education. In terms of Josiah Jr at least, the move had the desired result. Sarah was desperately proud of his academic achievements.

The behaviour of many of the gold-seekers she had encountered often stretched Sarah's levels of tolerance to their limits – and she was sometimes guilty of a degree of snobbery when she wrote about them – but her own standards never slipped.

'Any newcomer into San Francisco in those days had but to seek, in the right way, for good people, and he could find them,' she wrote. 'But in the immense crowds flocking hither from all parts of the world there were many of the worst classes, bent upon getting gold at all hazards, and if possible without work. These were constantly lying in wait, as tempters of the weak.

'Never was there a better opportunity for demonstrating the power and truth of Christian principle, than was, in those days, open to every faithful soul; and never, perhaps, were there in modern, civilized society more specious temptations to laxity of conduct.'

If it could be said of any of our Argonauts that they found their El Dorado, it would be Sarah. She died happy, righteous and fulfilled, in 1891.

When John Borthwick departed the goldfields in 1854, he travelled to Australia before returning to Edinburgh and publishing *Gold Rush: Three Years in California* in 1857. The book was well-received and his paintings were exhibited in several major galleries, including the Royal Academy of Arts in London. His accounts, both written and drawn, are among the most vibrant and historically relevant of all the literature and artwork spawned by the Gold Rush. He spent his last years in Paddington, west London, where he died in 1892.

William Swain arrived in New York via Panama in January 1851 to be met by his brother, George, who described him in a letter to Sabrina thus: 'He looked pretty thin in flesh, dark in colour and shabby in dress, and taken by and large was a hard-looking customer.' On 6 February the brothers arrived in Youngstown and almost everyone who lived there turned out to greet them.

'I have been many miles and seen many places, but this is the finest sight I have ever seen,' said William to George as they arrived home. He had been away twenty-two months. He brought with him a black satin dress from New York for Sabrina and, for many years afterwards, she would celebrate 6 February with William and wear it. The Swains went on to have three more children who were treated to readings from William's Gold Rush diary on special occasions. William died at the family home in 1904, aged eighty-three. Sabrina survived him by eight years.

After buying his passage home, William had returned with five hundred dollars' worth of gold and an interest in a claim that might have been worth two hundred more. It was not much for the trials he had endured, but he did not seem to mind once he became reacquainted with young Eliza.

He never struck it rich, but he had seen the elephant.

As for James Wilson Marshall, you might be forgiven for thinking that his story had a momentous ending, given such prominence, standing on his grand plinth overlooking the American River, but sadly you would be wrong. Gold might have made him famous but it also ruined his life. Some accounts say that Marshall continued to work the saw mill at Coloma for three years until gold miners upstream diverted the river and left it powerless; others have him fleeing the town after a dispute over the way gold prospectors were treating their Native American labourers. Whichever is true, he certainly left the area for several years, probably to prospect – unsuccessfully – higher in the Sierra Nevada before returning to Coloma in 1857 and buying a cabin on a small plot of land.

He tried to grow fruit and to make wine commercially, but these ventures failed. What little money he had left was invested in the Grey Eagle gold mine at Kelsey, just a short distance away, but he made no money from that. In 1872, in recognition of his discovery of gold, the California Legislature awarded him a monthly pension. The award ran for a two-year period and was renewed in 1874 and 1876 but not 1878.

Marshall died penniless on 10 August 1885.

Perhaps from a sense of guilt, perhaps gratitude, some members of the Native Sons of the Golden West, a historical society founded in 1875, came together the year after his death and decided that the man who started the California Gold Rush deserved to be remembered. After five years of campaigning on his behalf, they managed to persuade the legislature to set aside nine thousand dollars to build some sort of monument.

In the end, they settled on a ten-foot-high bronze statue of Marshall on top of a thirty-one-foot-tall granite plinth, pointing to the exact spot where, on 24 January 1848, he found gold.

In all the time I spent digging for gold, I found only one piece that could, at a stretch, be called a nugget, and I gave it to Gary on my last day in the Bear River. He was sitting in a garden chair in the shallows, his feet resting in the cool water, a cigarette hanging from the corner of his mouth. Both his hands were occupied, one with his bucket, the other with his trowel, and this meant he could do nothing to stop me putting a little vial containing the lump of gold into his top pocket.

'What the . . .' he said.

'You have it. You'll put it to better use than me.'

Gary had slightly complicated living arrangements inasmuch as he shared a home with his ex-wife, Lori, but they were no longer together. His continued presence not only reflected the fact that he and Lori were still friends, but also that he wanted to stay close to his stepdaughter, Crystal, and her daughter, Elizabeth, who was nine. Gary loved them both as if they were his own blood but recently he had decided to move from Sacramento back to his home town of Grand Rapids, Michigan. The domestic arrangements had become too difficult. However, the thought of leaving Crystal and Elizabeth behind was breaking his heart.

The previous day, Gary had been fretting about his imminent departure and was talking about it relentlessly, sharing the load with me, and that was fine.

'I'm going to have a real nice piece of jewellery made for Elizabeth out of my gold,' he'd said. 'Then every time she looks at it she'll think of me.'

A few hours later, I found what Gary called 'the nugget', but what I might have described as a picker. Either way, as my partner pointed out, it had length and height and depth and was most definitely not a flake. More than anything, though, Gary thought it had a pretty shape.

I was staying at a motel in Colfax and that night I emptied out my gold – the first time I had done so in weeks – and wondered how there could be so little of it considering the success that I thought I had been enjoying. It now covered a wider area across the plate, though most of it was as thin as foil. It weighed less than one ounce; I am embarrassed to say how much less, but there was barely enough to cover the cost of a flight to London.

I picked up the nugget. Gary was right; it did have some pleasing substance. In the morning I would give it to him for Elizabeth.

I caught myself in the mirror looking at my gold like a miser in a mountain cave, hunched over it, frowning. I stood up.

I had lost weight and caught a tan. I had muscle definition for the first time in years, even a flat stomach. I found myself laughing, not recognizing the person in the glass. My room had a veranda and I went outside with a cold beer and sat on it, watching the sun slink behind distant trees. It was time to go home.

'You don't have to do this, you know,' said Gary, taking the nugget from his pocket and examining it with a broad smile.

'I know,' I said. I hadn't told him that I had decided to leave.

I emptied my sluice box and washed the trapped sand and gold from its matting for the last time, and sat down in one of Mike's holes. Like all his diggings, it was only three or four feet deep. Any deeper and the water in it would have drowned him. I poured the pay dirt into my pan and slowly began to wash it away, losing layer upon layer of black sand with each gentle rocking motion. I could do this now without thinking. The fact that I could do it without thinking of anything at all made the act almost soporific.

Slowly, colour emerged as the sand migrated from the pan into the water. It was a golden matrix of precious pixels and they spread and fanned out, sparkling in the afternoon sun.

313

How funny, I thought, that the last pan I washed should be the most productive.

'Gary,' I shouted. 'Take a look at this.'

Bear River Gary climbed out of the hole he was working and made his way across the bar. I put down the pan and reached into my rucksack for a suction bottle to gather up the day's bounty.

'God damn!' he said, leaning over me. 'You dig like an American.'

I looked up at him, laughing and shaking my head as I clipped the pan, knocking it sideways and into the hole. I caught it at the perpendicular but all the gold had fallen out and I could see it cartwheeling through the water. For a moment I considered trying to grab at it desperately, foolishly, but it was too beautiful. I found myself hypnotized by a thousand flashes as the flakes and dust strobed through muddy shafts of light.

I pulled back my arm. 'Let it go,' I thought, and watched it vanish into the gloom.

POSTSCRIPT

The Luckiest Man in
the World

Tom Henderson underwent surgery and a course of che-
motherapy when he returned to Australia. At the time of
writing, his cancer is in remission.

Statue of James Wilson Marshall pointing to the spot where he found the first gold of the California Gold Rush

ACKNOWLEDGEMENTS

I would like to thank all the miners who put down their tools to help me and pass on their knowledge when it would have been much easier, and more fun, to point at me and laugh. Some of them are named in this book – Tom Henderson, Terry and JoAnne McClure, Duane Wilburn, Matt and Shannon from Mariposa, Heather Willis, Mike Morgan, Doug and Lonnie McDowall and, of course, Gene and Cathy Meyers. Others, such as Todd Osborne, one of the cleverest miners I met, are not. My deepest gratitude goes to you and to the wider gold prospecting community; you were kind to me.

My 'gold mentor' Nathaniel Burson was particularly generous with his time and to him I wish boundless luck. And how can I ever thank 'Bear River Gary' Shaver for teaching me to dig like an American?

Without the dogged support of my wonderful agent Laura Longrigg this story would never have been told. Huge thanks go to Laura and to Mike Harpley at Oneworld for his exceptional editing and wise advice, and to Paul Nash and Amanda Dackombe for spotting my mistakes and putting them right.

ACKNOWLEDGEMENTS

Heartfelt gratitude goes to The Authors' Foundation whose 2012 assessors Simon Brett, Sameer Rahim, Fiona Sampson, Helen Simpson and Frances Wilson awarded me a grant to help me write this book.

For giving me carte blanche to prospect on one hundred miles of claims, I would like to show my appreciation to Dave Mack, Rich Krimm and the New 49ers. For making me feel so welcome in Happy Camp, thanks go to Rita and Gary King, and their friend Beth Buchanan. I found Hazel Davis Gendron's peerless knowledge of the Karuk invaluable; thank you Hazel, and thank you Ellen Johnson who, with husband Bill, made me most welcome. And I simply must show my gratitude to Jeff Herman at Empire Mine for introducing me to *the guy*.

No set of acknowledgements relating to the California Gold Rush would be complete without a mention of E Clampus Vitus, whose ubiquitous guides and plaques were a constant source of information. I am secretly hoping they make me an honorary member.

At the Bancroft Library, University of California, Berkeley, I would like to thank Crystal Miles for her help and advice regarding Sarah Royce's *Reminiscences* and Maria Brandt for additional research. On the opposite side of the USA, I would like to acknowledge the generous advice and time afforded me by George Miles, William Robertson Coe Curator of the Yale Collection of Western Americana at the Beinecke Rare Book & Manuscript Library.

For giving of their time and their invaluable advice as readers, my gratitude goes to Andy McLintock – whose guidance on gold finance and economics was a life-saver – to Jan Owen, Bill Scannell and Jonathan Callery. For their support, my love and thanks go my mother, Pat, sister, Marie, nieces Nicola and Laura, nephew, Ben, and great niece, little Elsie, who is wearing some of my gold around her neck thanks to the jeweller Joanna

Pearce at MaisyPlum. An extra big thumbs-up goes to Ian Buchanan, whose strength has been an inspiration.

Peter Fearon, Emeritus Professor of Modern Economic and Social History at the University of Leicester, and Chris Blackhurst of *The Independent* were kind enough to check for errors in the chapter on the gold standard. If any remain, they are mine.

For their support and friendship, heartfelt thanks go to Mike Dolan, Steve and Andrea Redmond, Paul Hackett, Josie Martin, David Felton, Domenico Pugliese, Matt Elgood, Paul Peachey, Vanessa Thorpe, Sophie Woods, Mark Willis, Ian McLeish, Lindsay Frankel, Deborah Reade, Michele Callery, the Hallam family, Susan Speller, Zita Nicolaou Chen, David Lister, Douglas Maggs, Ruth Fielding, Pete Norman, Marie-Pierre Darneau, Louise Jury, Kathy Marks, Andrew Marks, Kirsty Bennett, Ben Unwin and John Hardwick.

And for having endless patience with me and all my daft ideas, my love and deepest thanks go to my wife and best friend, Suzanne, who is worth more than a gazillion times her weight in gold.

PERMISSIONS

Extracts from Sarah Royce's *Reminiscences*, (BANC MSS 72/53 c), reproduced by kind permission of the Bancroft Library, University of California, Berkeley.

Extracts from William Swain's diary and letters, reproduced by kind permission of the Yale Collection of Western Americana, Beinecke Rare Book & Manuscript Library.

Extracts from *Karuk, The Upriver People*, reproduced by kind permission of Maureen Bell and Naturegraph Publishers.

Extracts from *Hard Road West: History and Geology Along the Gold Rush Trail* (University of Chicago Press), reproduced by kind permission of Keith Heyer Meldahl.

Extracts from *The Power of Gold: The History of an Obsession* by Peter L. Bernstein (Copyright © 2000 by Peter L. Bernstein), reproduced by kind permission of John Wiley & Sons Ltd.

Extracts from *'Dear Charlie' Letters*, reproduced by kind permission of the Mariposa Museum & History Center.

Extract from *Gold: The Race for the World's Most Seductive Metal* (Simon & Schuster), reproduced by kind permission of Matthew Hart.